A Man Beyond Time

ROD NORVILLE

©2009 Nartea Publishing
a div. of DNA Press, LLC

A MAN BEYOND TIME
©Copyright by Rod Norville
©2009 Nartea Publishing, a div. of DNA Press, LLC. All rights reserved.

Library of Congress Cataloging-in-Publication Data

Norville, Rod. A Man Beyond Time / Rod Norville. — 1st ed.
    p. cm.

 ISBN-13: 978-1-933255-49-1 (alk. paper)

ISBN-10: 1-933255-49-8 (alk. paper)

1. Time travel—Fiction. 2. Black holes (Astronomy)—Fiction. I. Title.
        PS3614.O7828M36 2009
        813'.6—dc22

        2 0 0 9 0 1 8 9 7 2

DNA Press, LLC/Nartea Publishing
www.dnapress.com
editors@dnapress.com

Publisher: Nartea Publishing
Executive Editor: Alexander Kuklin, Ph.D.
Copy Editor: Rita Flórez
Art Direction: Alex Nartea
Cover Art: Studio-N-Vision (www.studionvision.com)

To my beautiful wife and soul mate, Roxanne.

Table of Contents

Chapter 1

DECEMBER, 1953 - THE FIRST DAY (Monday)

Except for a single black Ford, Riverside's Main Street was deserted on the hazy gray morning Greg Philips suddenly arrived disoriented and confused. He stood momentarily frozen. He had missed the planned midnight landing on the dry river bed outside of town. Recovering from that shock, Greg was slow to see a car careening and fishtailing toward him from a small cross street a few yards away. The driver swerved at the last instant, grazing him and sending him into a pirouette to his knees against the curb.

"Sonuvabitch!" Greg grunted.

The car skidded violently to the left across the street. The driver expertly turned the steering wheel into the slide while pumping the brakes. With a screech of rubber, the car fishtailed back to the right and slammed over the curb onto the sidewalk. It settled to a stop with its right front wheel rim crushed and tire blown. The passenger door flew open and a tall, skinny man jumped out.

Greg was bruised and dazed. He slowly pushed himself to his feet and watched the driver, who ran around the rear of the car and joined his passenger inspecting the damage. They were streetwise looking men in their late teens, dressed in jeans and scruffy leather jackets. The passenger was well over six feet tall, with hunched shoulders and had a graceful way of moving with a minimum of effort. The driver was a few inches shorter, with a thick shock of light brown temple hair currently in disarray. Greg recognized them and went weak in the knees. He saw an alley a few feet away and awkwardly scurried into it to avoid a compromising encounter.

"Hey, you!" the driver yelled.

Greg surged down the alley. In the early morning silence, he heard their footsteps accelerate and pound after him. The narrow alley branched into a T, and he lurched to the right. There was a dirty stucco wall on his left and a stainless steel door above a black-grilled, three-step porch to his right. The door was held open by a mop-filled bucket. Greg climbed the steps and dove through the door, slamming it behind him. He leaned against it, listening with apprehension to the sound of footsteps hammering past in the alley.

Greg thought, "How in hell did he find me the minute I arrived? What's going on? That's what Ken said I had to avoid at all costs. Damn!"

As his eyes became accustomed to the darkened interior, Greg realized he was in an empty, large and elegantly-appointed movie theater. He had entered through a fire exit halfway between the screen and lobby and could hear the muffled hum of a vacuum cleaner seeping through the heavy vermilion curtains separating the two. He unhooked and removed his restrictive suit and stashed it under the deeply-padded leather loge seats. Now dressed in tan khaki trousers and red-checkered wool jacket, Greg walked up the aisle with military posture and opened the lobby curtain.

A doughy, middle-aged Mexican woman pushed an industrial-size vacuum cleaner across the lobby carpet. Greg shoved open one of the large glass front doors and walked boldly across the alcove past the empty ticket booth.

He looked in both directions as he stepped out of the alcove onto the sidewalk. At one end of the block his two pursuers were pushing the disabled Ford into a parking space. Adrenaline rushed through him, and he leaped back. The sudden movement was a mistake.

"Hey, you!"

Greg spun and dove back into the lobby and darted up a broad stair-case to the balcony. The startled Mexican woman's mouth dropped open, and she switched off the vacuum. As Greg moved out of her sight, the two angry and wild-looking young men raced into the lobby, stopped, and looked about. The driver waved his companion toward the balcony, shoved the main theater curtains aside, and entered. Greg ran lightly down a balcony aisle toward the screen, turned into a cross aisle and saw an illuminated emergency exit door. He heard the muf-fled clumping of someone bounding up the staircase behind him and realized he'd never make it to the door before the pursuer would be upon him. Greg lunged between seats a row from the door and ducked below seat height.

"See him?" his pursuer yelled to the driver below.

"No, but he's here. The prick didn't have time to get out. Look between the seats!"

The familiar sound of the driver's voice echoing in the empty the-ater swept Greg with an arctic blast of fear. He raised his head slowly and peered between the seats in the low level lighting. The taller one, all bony edges and angles, walked quickly up the aisle, leaning over and looking between each row before moving on. Greg lowered his head. In a minute the pursuer would find him. Sweat soaked his jacket. He rolled onto his back and pulled his knees into his chest. The tall man yelled to the driver below, "I hear him!" He vaulted up the aisle, stopping two rows shy of Greg. Greg saw his feet move closer to the next row and braced himself. As the young man's face leaned into his row, Greg kicked out both feet. The pursuer grunted and tumbled onto his back, blood spewing from his nose.

Greg scrambled over the seat and dove for the exit door. He heard the driver charging up the balcony stairs. Fear tore through him. He was through the door and looking down a metal staircase with a wrought iron railing. Greg knew he'd never get down it before being

3

overtaken by the much younger man. He slipped behind the door and grasped its metal handle with both hands. As the young man ran into the opening, Greg slammed the door in his face, bounded down the steps and ran through a narrow passageway between two buildings and out into the street.

The street was familiar. Greg remembered there was a hotel a few blocks north and ran toward it. No cars were out, and not another person was in sight. Fear pried into him at the thought of the consequences had the driver reached him and drove him on, badly winded, the muscles of his thighs and calves burning with pain. As Greg approached the hotel entry, he slowed to a walk and looked apprehensively over his shoulder. The hangover headache that had been resting just behind his eyes made a throbbing reappearance.

Once checked in and settled into his room, Greg found himself trembling. He sat on the bed, rubbed his bruised hip and tried to look at the morning's events with scientific objectivity. The whole event was a fluke, he thought. The young man hadn't recognized him, he'd chased him because he was ticked-off about his car. He arrived in the wrong place at the wrong time. That's all. He lay back, forcing himself to breathe slower and noted the time on the room's clock.

At a few minutes after 9 a.m., Greg got up, washed his face and left the jacket on the bed, reasoning the young men had only briefly seen a figure in the distance wearing a red-checkered jacket. The sidewalks should now be filling with shoppers, and without the jacket, he should not stand out. He rubbed his sweaty palms on a bathroom towel and exited the hotel.

Chapter 2

The old man shuffled along the gravel river bottom road several blocks from his home as he had every weekday morning for years. He found it a pleasant way to start the day. First coffee and a donut with his wife while reading the morning Enterprise and then his ritual 40-minute walk from his home through open fields into the Santa Ana river bottom before opening his shop precisely at 9:30 a.m. The air this day was pale and soft and the landscape undulated with hues of pale green and brown.

It was good here in the quiet morning with the smells of damp earth and rich decaying compost. A light breeze came up as he left the gravel road and walked along the bank of the riverbed. Now in December, there was a little water, and it cleaned the air that moved coolly down the river basin. A few feet from the path was a profuse variety of brush and trees. For a long stretch, he walked beneath a canopy of sycamores. A radiant and rosy light filtered through the undergrowth.

He nurtured no hopes or future dreams. He had nothing left undone, only his memories and the occasional fear of what might be in store for him after death. He enjoyed remembering the eagerness and probity of his youth and thinking the good thoughts about his wife. Late in life, she had brought him a peace and tranquility he had not known before.

He reached the turning point of his walk and stopped to pull his chained watch from his pocket. It wasn't necessary because the time never varied more than a minute at this point. As he closed the cover and pushed the watch deep into his front pocket, he suddenly had a feeling someone or something was watching.

He heard no sound and saw no movement except his own. But a

sense, rarely used, told him he was not alone. He pushed the feeling aside, deciding that his senses were old and undependable. Moving on, he thought he saw movement deep in the brush in his peripheral vision. He stopped and squinted toward the now still undergrowth.

"Who is it?" he called.

When there was no answer, he decided his old eyes were unreliable, as well. He walked on. Something was different, though. The woods had become abnormally silent. He stopped again and listened. There was a faint rustle of dry brush. A cold shiver washed up his back.

Something — or someone — was out there.

He'd not heard of a raccoon or coyote in these parts for years. There were no animals in this river bottom large enough to harm him. He moved on in confidence. Suddenly, he heard something crashing through the thicket.

"Who's there?" he cried again.

He began an old man's frightened trot toward the gravel road. Something ran parallel to him in the brush. Something big and fast. Whatever it might be was trying to reach a point before the gravel road, where the brush met the water and where they would momentarily be out of sight of any eyes. He ran faster, but his legs threatened to buckle under him. His mouth was dry, and his breathing labored as he struggled to beat this thing to the road and the security of a passing car.

The thrashing stopped. A large dignified-looking man stepped gracefully onto the path a hundred feet ahead. The man casually pulled a pack of Old Gold Cigarettes out of his shirt pocket and lit one with a burnished, gold lighter. The old man froze in his tracks, gasping for air, and observed this stalker blocking his way. The stalker waved him forward, smoke curling around his smiling face.

Chapter 3

Greg left the hotel, paused, and ran fingers through light brown hair going gray at the temples. He was slim, an inch under six feet, and walked with the self-confident ease of a man who'd practiced 15 minutes of martial art exercises every morning for years. After looking around to get his bearings, Greg admired the ambiance of the town, a welcome oasis of citrus trees and irrigation canals bordering the great Colorado desert that straddles Arizona and California. Riverside was so named because it sits beside the seasonally dry Santa Ana River that flows west into the Pacific Ocean.

Greg walked east to Main Street and entered Citizens Bank. He filled out a new checking account form, listing his name and Riverside as his birthplace 48 years ago. He logged his address as the prestigious De Anza Hotel, noted his social security number, and deposited $2,000 in $100 bills. He received an account number and several temporary checks.

Mixing with the morning shoppers, he walked north two blocks to Hogles Brokerage, a spartan regional stock brokerage house. Early morning investors sat on hard wooden benches against the wall, their cigar and cigarette smoke rapidly filling the room. They concentrated on the ticker tape stock prices running past on the opposing wall. The tape was projected from an adjacent room onto a mirror and then reflected onto a silk screen in the main room. Its focus quality reminded Greg of a blurred carbon copy.

But he had no interest in these "Big Board" stocks and approached the floor broker, a frail little man with a choirboy face in his early thirties.

"Hi, my name's Greg Philips. I just opened a new account."

The room was oppressively warm and smelled faintly of stale tobacco smoke.

"Pleased to meet you." the broker beamed and squeezed his hands together. "What can I do for you?"

"I'm interested in buying shares of Cinerama Inc." Greg replied.

"Doesn't ring a bell. New York or American Exchange?"

"Neither. It's over-the-counter."

"Just a minute ... here it is. It's a penny stock. Looks like it's been pretty volatile."

The broker moved to a heavy oak desk and filled out a quote request form. He stuffed it into a cylindrical container and shoved it into a vertical vacuum tube. He turned and saw Greg shifting his weight from one foot to the other. He squeezed his hands again and said, "It'll take a few minutes. The wire operator has to teletype a 'Bid and Ask' request to Cinerama's O.T.C. trader. Why don't you take a seat?"

Greg joined the other investors staring at the wall and awaited the return teletype. Ten minutes later, the broker beckoned and handed him a piece of paper. "Here it is. An asking price of 40 cents per share."

"I'll take 5,000 shares," Greg said and filled out one of his temporary Citizens Bank checks. "I'm new at this. If I wanted, could I sell this stock tomorrow and get paid?"

"Sorry," the broker's beaming countenance dimmed. "We're governed by the New York Exchange Statute 405. It requires an incubation period of five days before funds — of course less commission — can be paid. However, you can reinvest it all immediately including profit in other stocks if you'd like."

Greg nodded a thank you and walked back out onto the street. He proceeded another block east to his fourth stop, Hertz Car Rental. Greg showed the male clerk his driver's license and rented a late model Chevrolet. An old German Shepherd, with hips going bad, was lying by the chain-link fence next to where the Chevrolet was parked. The dog stood up and wagged his tail. Greg walked over and scratched his ears.

He drove out into midday traffic. Sounds were muted to his ears, and he felt slightly nauseous. Greg pinched his nose and blew to equalize his ear pressure. He assumed his nausea was from his trip, hangover, or both.

He parked in front of Mape's Cafeteria. After navigating the scarred wooden tables to the glass-covered food line, he studied the wax paper covered sandwiches. Greg had a penchant for tuna and selected one on whole wheat toast. A petite, bespectacled brunette in a tightly fitted starched white uniform, a cotton cap pinned saucily to her hair, approached with a steaming pot.

"Cup-a-coffee?"

Greg smelled the aroma of freshly brewed coffee and answered, "You bet. Thanks."

She turned his cup upright in its saucer and was about to pour when she looked into his face. She paused and licked her lips and radiated a return smile as she poured. "Will I be able to get you anything else?"

"Change these for nickels?" He pulled two single bills out of his pocket.

"Coming right up." She turned and gave her behind a practiced seductive roll as she walked to the cash register. Greg caught a long forgotten scent of gardenias. That's the perfume Janet had worn, he thought.

Dim and silvery memories were suddenly stirred as he slowly unwrapped his sandwich and stared down into his steaming coffee. On a typical Friday night 30 years ago, he had pulled into the furthest row from the screen of the Rubidoux Drive-In theater, casually steering with his left hand on a necker knob while his right circled Janet Ward's slender frame and cupped her right breast.

"I'll get us a couple of sodas." He had lit a cigarette with a Zippo lighter and swung his door open.

"Wait, I'll go with you. I'm going to the ladies room."

Greg had stopped at the soda fountain and watched Janet sashay on to the rest room, her derriere firmly ensconced in a belly to thigh girdle beneath a tightly fitted knit skirt. Her breasts were thrust unnaturally up and out by an equally constraining metal supported brassiere underneath an angora sweater. He bought two Cherry Cokes and returned to his car. He flicked his cigarette into the night when he saw Janet flouncing back comfortably jiggling front and rear without restraint - a handbag overflowing with her undergarments.

Greg sat the Cokes on an open glove box door as she slid into his arms. Her tongue explored the inside of his mouth.

She pulled back and asked huskily, "Back seat?"

Minutes later, Janet's feet were marking up the ceiling upholstery. The windows were steamed up all around and the car was filled with the sea-brine smell of sex. Janet's moans were drowned out by the concomitant sounds of gun-fire from John Wayne's latest western.

The waitress laid a handful of nickels on the table. "Some more coffee?"

"No thanks." Greg pushed forward a generous tip and pocketed the nickels.

"Sure I can't do anything else for you?" She placed her hand over one of his as he raised out of his chair.

Greg smiled and shook his head. The sun, this brief December day, shone weakly in the sky, and he shivered. After pausing by his car to stuff the meter with nickels, Greg stopped into Sears and Roebuck and purchased an off-the-shelf gray wool sports jacket.

The dissonance of chimes a block away announced the hour. At the corner, Greg saw their source, the grand old Mission Inn. Its rococo architecture overflowed into a labyrinth across an entire city block. Its gardens, a runaway luxuriance, were scrupulously manicured and trimmed. A feeling of severe homesickness swept over him.

Greg remembered the necessary brush-off he had given Janet on his last day in Riverside 30 years ago. They had stood polarized in the crisp evening air awaiting his departure on a Greyhound bus heading north. The day had been overcast and dark.

"Why are you doing this? The Korean war is over." Janet had stood before him twisting the fabric of her corduroy handbag. The Greyhound bus, engine idling, steamed in the winter evening air. Greg stood awkwardly, using both hands to hold a battered suitcase.

"What do your parents say?" Janet wiped the back of her hand across her nose smudging her lipstick.

"They could care less."

"Is it something I've done wrong?" Janet grasped at his sleeve. "Do you think I've been fooling around on you?"

"No." Greg gently removed her hand. "It's time for me to get on board." His voice cracked and sounded strange to him. Greg climbed into the bus and sat in a window seat. Janet walked along side the bus as it pulled into the street. She was clutching her handbag closely to her breast, her face contorted. She couldn't see through the fogged window that his eyes were also brimming. Greg withdrew from the other Air Force recruits into the inky sanctum of the rear of the bus. "I have to run, Janet," he inaudibly cried.

Two years later, while stationed at a base in Savannah, Georgia, his mother sent him a newspaper clipping announcing Janet's marriage to his friend, Ralph Gould. Thirty days later, Greg married a Savannah girl named Helen.

Greg turned away from the Mission and walked on to his next appointment.

AT THE RIVER

The stalker nonchalantly lifted the old man off the ground by his neck and dragged him into the trees. He then slapped him repeatedly across the face before asking in an aloof but not unfriendly voice, "Where is he?"

"Who?" the old man wheezed, tears stinging his eyes.

The stalker lowered the old man's feet to the ground and repeated, "Where is he?"

The old man smelled the faint scent of an expensive aftershave. He answered, "I don't know what you're talking about."

The stalker dispassionately removed the cigarette from his lips and ground it out on the age-marked forehead. The old man mashed his teeth together and gasped, but refused to cry out.

The stalker dropped him onto his back and placed a knee onto the bony breast. He slowly took an arthritic hand in his own. The old man watched him — mesmerized. A finger was selected. "Where is he? I won't ask again."

"Who in hell are you talking about?"

"Your son."

"I haven't seen him in a couple of days."

The stalker slowly bent the finger backward until it snapped.

Chapter 4

G reg entered the public library where he skimmed through recent copies of both the *Morning Enterprise* and the *Evening Press* newspapers. After 20 minutes, he found what he was looking for buried in the local news.

Rival Riverside and San Bernardino gangs clashed again after Friday night's football game between the Riverside's Bears and the San Bernardino's Wolverines at Polytechnic High. Police broke up the altercation, but not before an 18-year-old-San Bernardino youth was injured. Antonio Romo was taken to Community Hospital emergency with a severe wrist injury... .

Greg knew that had not been the end of it. It had led to a $100 drag race bet between the gangs to maintain their honor. Greg's hands trembled, and the scene of what had made him run 30 years ago replaced the newspaper before him:

On a moonlit night a few months after his high school graduation, two souped up V-8 engines idled side by side on an asphalt orange grove access road. Romo at the wheel of a three-window blue '39 Mercury coupe — his left foot on the clutch, right foot poised above the gas pedal, left hand on the steering wheel while his bandaged right hand gripped a small metal skull screwed onto the end of the steering column mounted stick shift currently in first gear. A large pair of fabric white dice with black dots hung from the rear view mirror. Young Greg was at the wheel of his '40 Ford grey primer sprayed two-door sedan. It sat low to the ground in the rear with six-inch-extension shackles. Like Romo, Greg was poised to pop his clutch and slam his right foot to the floor. His left hand gripped the necker knob, and his right gripped a sawed off floor mounted stick shift. Angry and agitated young men surrounded both cars. Ralph Gould, in faded levis pulled down low on skinny hips, stood facing them a dozen yards in front. He held a raised flashlight pointed to the sky.

Ralph slammed the flashlight down between his legs and Greg stomped his throttle to the floor. The Ford's dual Smitty mufflers roared, and Greg smelled scorched rubber from the rear tires. The two vehicles careened past either side of Ralph as Greg slammed his stick shift into second and popped the clutch again with a screech of second gear rubber. Romo pulled ahead but missed his speed shift into third and fell behind, his left front fender next to Greg's right rear fender. Romo tried to force Greg to the left as they approached a single lane bridge. Seconds later there was the roar of a crash and an explosion.

Greg set the newspaper aside, leaned back and silently mouthed, "Antonio Romo." He found a San Bernardino telephone book, but there was no listing for a Romo family. He walked out into the warming afternoon sun and headed toward Brunlon Optical.

An elderly optician dressed in an oversized, off-white smock tested Greg's sight.

"Don't see astigmatism this severe every day." His voice was whiny.

"I lost my glasses. When can I expect to get these?"

"About a week."

"A week! I'll pay double for a rush order."

"I'll call one of my regulars and see if he'd mind waiting to help somebody with a similar dysfunction. Come back on Wednesday."

Greg looked at the clock. It was 3:00 p.m. He had one planned stop left for the day. He returned to his car and added the remainder of his nickels into the meter and entered the Chi-Chi Lounge on the corner. He settled on a stool at the bar.

"What'll it be?" A heavy set bartender in his mid-40s was wiping down the counter. The darkened bar's maroon rug smelled of decades of spilled drinks. "Southern Comfort Manhattan. On the sweet side." The bartender, with a bald pate and surrounding fringe of graying hair, mixed the Manhattan and dropped in a maraschino cherry. "How about all the excitement around here this morning?"

"What excitement?"

"Big time excitement!" The bartender resumed idly wiping the counter. "A couple'a punks chased a guy into the Fox Riverside theater right across the street. The Mex cleaning lady saw the whole thing. Must'a had a helluva fight in the balcony. The cops found blood all over the place."

"Any arrests?" Greg took a sip of his drink.

"Naw. The cops figured they all beat it down the back stairs. There were blood drops there, too." He waved the wet towel in the general direction of the theater.

"How do you know about it?"

"Saw the cop cars in front as I was coming into work. Seems the Mex broad ran outside when she heard the ruckus and waved one down."

"What do the police think happened?"

"Don't know, but they sure found something strange." The bartender shoved forward a bowl of peanuts. "While I was talking to a cop out front, his buddy brought out what looked to me ... like maybe ... hell, I'm not sure. Just some of the strangest looking clothing I ever saw!" He raised two pudgy hands. "I kid you not."

He grinned and moved down the bar to one of the regulars who had just entered. "Hey, quite the excitement around here this morning, eh?" he repeated as he got the regular his beer. After a time, he returned to Greg. "New to town? Haven't seen you before."

Greg thoughtfully pulled on his earlobe. "I used to live here a long time ago. I've only been back two times in 30 years." Greg downed his drink.

"Want another?"

"Better not."

He ambled down Main Street and looked carefully at the pedestrians fearing and half expecting to encounter someone he'd known. It was a little after four when he stopped to look down an arroyo overlooking his old high school's playing field. After school football practice was in progress, the players wore their familiar blue and gold jerseys. Greg was jerked from his thoughts by a voice softly addressing him from a few feet away.

"We do have an excellent team this year, don't we?"

A small, slender woman dressed in a fashionably severe suit stood a few feet away. One hand held the railing as she looked down upon the play below.

"If only they could transfer some of their enthusiasm for football to academics," she said.

Greg caught his breath as he recognized Emily Johnson, his physics teacher so many years ago. He took a moment to compose himself and said, "You must teach here."

She wore a blue pinpoint button down shirt beneath the suit, and carried a worn leather briefcase almost overflowing.

"Physics," she looked at him. "Do you have a boy down there?"

"No, I don't live here," Greg paused and said. "I'm just in town for a few days."

She waved her free hand toward the palm trees lining the sidewalk. "It's a nice little town, isn't it? Big enough to have a little culture, but small enough that I can walk to school."

She waited for a truck with a noisy exhaust to pass and studied his face. Her blue eyes narrowed. "Are you here on business or pleasure?"

"A little of both."

She set her briefcase down and put out her hand in a formal way. "I'm Emily."

"I'm ... . Some people call me Phil." He shook her hand.

"I'm pleased to meet you, Phil. Is this your first visit?" She raised her voice to avoid being drowned out by sudden yelling from the players below.

"I was here years ago. I live in Menlo Park now. A small town south of San Francisco."

"I've heard of it. Has Riverside changed much since you were here last?" She studied his face.

Greg nodded toward a small mountain a dozen blocks away. "It's what I remember. Mount Rubidoux still looms over town and, of course, everywhere I look I see orange trees and soot."

A familiar smell of soap and perfume drifted toward him.

"Memories get selective as we grow older," Emily agreed, "and we'll have the soot as long as we have smudge pots to keep the oranges from freezing on these cold winter mornings." She smoothed down her skirt and picked up her briefcase.

"How about you, Emily, are you a native?"

"No, I'm a farmer's daughter from Blythe. You look familiar. Do you have family here?"

"No."

"That's interesting. You look a lot like one of my students from last year," Emily said. "Well, you'll have to excuse me. I have papers to grade. Have a pleasant evening."

"You too, Emily."

Greg was flushed with excitement as he watched her move away. He remembered the roseate Emily Johnson and the impact she had on him in her physics class. His life-long love of science had been an inestimable gift from her. She had been in her 20s and a dedicated teacher. After one of their confrontations, she had kept him after class; her anger had been measured in the slow burning of her checks.

"You're wasting your time and mine with your childish behavior." Greg played with the cigarette perched behind his ear as she continued, "You're smart enough, but going nowhere. You're an irresponsible brat."

Greg fell in love.

Emily was the quintessence of womanhood. He'd buckled down to impress her and watched in rapture as her diminutive figure paced the classroom stabbing the air with a piece of chalk as she lectured. She had succeeded where his parents had failed in setting an example for him during his final year in high school. He realized Emily had never been far from his thoughts for 30 years.

Greg headed back in the direction of his car down Market Street to the Press-Enterprise newspaper building, where he'd make his last stop of the day. He entered and placed an order for a paragraph to be printed in the next week's personals. *"Ken, old buddy, I'm alive and well in 1953. Give my best to Pee Bee, Greg."* At Owl Drugs he purchased a Gillette single blade razor, toothbrush, had a light dinner in a small cafe and returned at dusk to his hotel.

The night clerk had closed the small front desk and dimmed the lobby lights, so it took Greg a moment to observe a figure settled in a rocking chair under the staircase. The man appeared to be in his early 40s, big with well developed shoulders. His baldness reached up from his temples to a little promontory of hair. He was dressed in an expensive well-tailored suit several years out of fashion. A cigarette was burning to ash between his fingers. A pack of Old Golds lay nearby. He stared at Greg with a malevolent intensity.

Disconcerted, Greg scurried from the lobby, through the Spanish adobe tile courtyard to his room. He double-locked himself in.

Chapter 5

THE SECOND DAY (Tuesday)

G reg awakened refreshed and in good spirits. After a shower and shave, he gathered his Chevrolet from the parking lot, drove to a small cafe and ordered coffee, a sweet roll and a morning Enterprise.

He was reading by a large plate glass window overlooking the street when he heard the cracking thunder of a car's dual exhaust system. The Ford that had nearly killed him the previous morning sped by. Greg dropped the paper onto an adjacent chair, and waved to his waitress for a check.

He parked across the street from the brokerage house and waited until the lobby had filled with investors. He placed an immediate sell order with the eager floor broker on the Cinerama stock he had purchased the previous day. During his wait with the coterie of smoking investors, the wire operator teletyped Cinerama's O.T.C. trader. The return teletype confirmed the volatile penny stock had soared from 40 cents a share to a $1.20. Greg promptly reinvested the $6,000 by placing market orders with the floor broker for stock in two other penny stocks: Gaspe Oil Venture and Nevada Tungsten.

He then purchased a canvas suitcase, a change of off-the-shelf clothing, five pairs of underwear, an inexpensive wristwatch, and a small tool kit at JC Penney.

His next stop was Reids Automotive Yard at the corner of 10th and Lime. A string of barbed wire covered the security fence. Greg strolled through the derelicts, not finding what he was looking for and spent the rest of the day visiting yards in adjacent towns. He found the first item he needed in San Bernardino and the second in a yard in Colton.

It was nearly four in the afternoon when Greg returned to Riverside, keeping a constant eye out for the Ford. He wanted to see Emily again. The anticipation of it leaped up bright and unexpected in his mind. In memory, he saw her laughing at his youthful anecdotes. Her appeal had been as much in the way she walked and carried herself, as it was her beauty: her voice, soft, earthy, feminine, yet unaffected and with a note of strength and the way her eyes would meet his, forthright and interested. He cruised by the high school playing field hoping she made it a ritual to stop there on her way home.

She was there. He pulled over, rolled down the window and called out to her.

"Oh, hello," Emily said.

"Need a lift?"

Emily looked at him intently and nodded, her lips pursed. She opened the door and climbed in. "How'd your business go today?"

Greg pulled away from the curb. "Fine, thanks."

"Does that mean you'll be leaving?"

"As a matter of fact ... I'm considering moving here. Could I interest you in accompanying me in a look around?" He turned his head briefly to look at her.

"We are friendly trusting people here in Riverside. If it doesn't take too long, I'll be pleased to help you get reacquainted."

"Great! Let's start with a view from the top of Mount Rubidoux."

He drove west to the high pillars supporting a wrought-iron gated entrance to the mountain road and started the steep serpentine climb. Emily sat thoughtfully with her hands folded and her knees together.

"Who are you?" Her voice was somewhat demanding.

Greg whiplashed his head toward her. "What?"

Emily tapped her knee with a fist like a judge gaveling for order. "Phil, I saw my student today. The one you reminded me of. I had to confirm what I thought I saw yesterday. He looks like you. So much so, I believe you have to be related."

Greg leaned his head back and took a deep breath.

"I quizzed him about you," Emily continued. "He's never heard of you ... but you're the same height, build ... walk alike and sound alike. You must be related."

Greg concentrated on guiding the car round and round up the single lane road.

Emily watched his body language. "Would you like to meet him?"

"No, I can't say that I would."

Several minutes passed in uncomfortable silence. Emily's back straightened, and she turned in her seat, a determined set to her features. "I'm very fond of this young man. I have to be sure you aren't a problem for him. He doesn't have any relatives that I know of except his folks. He's sensitive and bright, and if you are a problem, I want to know it." She waited for a response that didn't come. "There isn't a doubt in my mind you're closely related. You are so alike he could be your son."

Greg pulled into a parking spot at the top of the mountain. He sat stiffly, looking straight ahead, his hands gripping the steering wheel so tightly that his knuckles were white. He turned slowly to Emily. "I can't tell you why I'm here, but I assure you I'm not here to make problems for anyone. In fact, I hope my visit will be good for several people."

Emily puzzled for a moment, then she spoke with measured emphasis, "Who...are...you?"

Greg opened her door, and they walked across the small parking lot and climbed the stone stairs to the massive Easter cross at the apex of the mountain. Greg looked west over the Santa Ana riverbed below and pondered for a minute.

"I'm a physicist. I do my research with electron microscopes."

"What's an electron microscope?" Emily crossed her arms over her breasts as a cold wind smelling of mesquite kicked up on the hilltop and ruffled her dress.

Greg shrugged off his jacket and put it around her shoulders. "It's a tool researchers use to magnify things up to a million times."

"Does your coming here have anything to do with your research?"

"Not really."

"Then, why are you here?"

Greg paused and carefully selected his words. "To make amends for something I did 30 years ago."

"Amends for what?"

"I'm sorry, Emily. I can't tell you that."

"Is it possible that my student is your ... son?"

"No."

"Are you married?"

Greg shook his head and carefully chose his words. Lying was anathema to him. "Not now. I once was."

"You told me yesterday you were from Menlo Park. Is that where your former wife is?"

"Yes."

"What does she do?"

"Nothing. She has annuities from a wealthy family."

Greg gently took her elbow and walked them back down the steps to the car. "I worked at the Stanford Research Institute in Menlo Park. I was fortunate to work for one of the finest scientists in the Country. Dr. Kenneth Hoard."

"Are you also a doctor?"

"Yes. I received my Ph.D. from Cal Tech."

"What is the Stanford Research Institute?" Emily took his hand and carefully selected her steps.

"SRI is a giant think tank that conducts research for government and private industry. Over the years they've made major advances in computer technology, genetics, and laser communications."

Emily glanced at him. "I'm not familiar with those technologies. Anyhow, I don't see any connection between -."

"Could we talk more about it tomorrow?" Greg interrupted and opened the car door. "It's been a long day."

"Of course! I want to get to the bottom of this. There may be a connection between you two you're unaware of. You're going to have to meet him and see for yourself. I'm free tomorrow after school."

Greg started the car. "I don't know what my seeing him would accomplish, but I'll pick you up tomorrow. Same time, same place. We can have an early dinner in the Mission Inn?"

Emily face softened and she said, "Okay, I live at the corner of Sixth and Walnut."

They sat in silence during the drive to her home. Greg pulled up in front of an old Victorian house with white clapboard sides and yellow and brown trim in a working class section. Two stories had been converted into a duplex with a big front porch with swings and rockers. It had better landscaping and maintenance than the other houses on the block.

Greg helped her out of the car and took both of her hands in his. "I'm only here to help people, Emily. I promise."

She freed one hand and fumbled in her purse for her keys. "I'm sensitive enough to recognize a good man when I meet one."

She stepped through the doorway, closing the door behind her.

Chapter 6

STANFORD RESEARCH INSTITUTE (SRI): SUMMER 1973

The electron microscopy laboratory main phone rang just as Greg was about to shut down the Hitachi Model HU11E electron microscope for the day. "Greg? It's Ken. Can you meet me for a drink tonight?"

"Sure. What's up? Anything special?"

"Special. I'm involved in getting another government contract for us and wanna talk to you about it."

"I'll give Helen a call and tell her I'll be late."

Greg drove the few blocks from the SRI parking lot to the Red Cottage restaurant and lounge. Once inside, he saw Ken Hoard sitting at the bar with a black, combination-locked briefcase in his lap. His middle-aged wrestler's body set on a short frame was easy to identify and, although he was now a division manager, he still wore a Texan string tie and blue western shirt. A bush of salt and pepper hair surrounded an intelligent and sensitive face. It broke into a crinkled eye smile when he saw Greg and he stood and grasped Greg's hand.

Ken had been amicably divorced a dozen years and lived a spartan existence in a studio apartment two blocks from SRI. He had two loves in his life, his science and his son. David was a mirror image of his father and had spent most weekends and summers with him. He was now in his senior year at the University of California at Berkeley as a computer science major. Ken was very proud.

They moved from the bar into the lounge and slid into a corner booth with a well-padded brown leather seat. The room smelled faintly of stale beer and pretzels.

"The night's on SRI, buddy. I've got a lot'a stuff to talk about," Ken said in an East Texas twang.

"Drinks on SRI? That'll be a first." Greg removed his wind breaker and laid it on the seat beside him. "What's so important?" He then leaned forward resting his elbows on the sticky plastic surface of the table.

"Let's start with Uri Geller," Ken said. "What do you think?"

"He's impressive." Greg looked up and smiled at an approaching waitress.

"Nice to see you, Molly," Ken said to the buxom waitress in her early 40s with too-black dyed hair. "How about a couple'a Anchor Steams?"

"Nice to see you guys." She wiped down the table with a wet towel. "Anchor steams coming right up."

Ken turned his attention back to Greg. "Is Geller for real?"

"I only met him a couple of times when he came to the EM lab with Russell Targ looking for scanning electron micrographs on some of the fractured surfaces of spoons and forks he'd presumably bent and broken without touching. My impression was he's sincere in trying to scientifically validate his telepathic abilities."

Greg paused. "I'm actually a little in awe. I saw him perform at the Flint Center last year. I was cynical but can't challenge what I saw. Members of the audience brought him knives, keys and broken clocks and held them out to him. He'd appear to concentrate and the clocks would start ticking and the knives and keys would start bending. If he's a charlatan, you couldn't prove it by me."

Ken nodded. "It's been that way wherever he goes. He got his first international attention a few months ago on the Jimmy Young TV show in England. He asked the viewing audience to join his concentration, and the telephones started ringing off the BBC walls. Electrical items of all types were turning themselves on and off all over the island. He did the same thing here, to a lesser degree, on Jack Parr's Tonight show and Merv Griffin's daytime spot."

Molly set a tray down in front of them and pushed their beers forward. "Here's some pretzels on the house, fellows."

Ken smiled. "Thanks, Molly." He turned back to Greg: "Tell me about the EM micrographs."

Greg chewed thoughtfully on his lower lip. "I've corroborated William Franklin's results from Kent State. The majority of the breaks Uri seems to induce do resemble fatigue fractures except for a platinum ring that spontaneously developed a fissure in its surface in Geller's presence — two breaks produced by two different conditions. One resembles a cleavage that occurs at extremely cold liquid nitrogen temperatures and the other at platinum's melting temperature ... Crazy? Yes, but you can't argue with an electron microscope."

Greg paused. "Why is SRI interested in him? Telepathic prowess hasn't exactly been our thing."

Ken drew little patterns with his index finger on his mug. "His psychokinetic ability to bend things is not what we're interested in, but your corroboration with Kent State firms up my decision we should go forward with him."

"Where do I fit in?" Greg sipped his beer.

"We've received a contract from the CIA to study out-of-body remote viewing, officially referred to as astral projections. Here, read this." Ken punched in the briefcase's combination, pulled a pamphlet out and pushed it across the table. "This is secret, so put your government Q clearance hat on."

Greg took brown horn-rimmed glasses from his shirt pocket and quickly scanned through several double-spaced typed pages. He carefully refolded his glasses and slowly slid the pamphlet back to Ken. "Out-of-body espionage? Astronaut Edgar Mitchell attempting to send

ESP signals from the moon to psychics on Earth while on his Apollo 14 mission? Wild stuff."

"What triggered the CIA interest was a speech by Soviet Premier Brezhnev earlier this year urging the U.S. to agree to ban research and development of weapons more terrifying than nuclear weapons." Ken put the papers back into his briefcase. "Every Western intelligence agency has tried to find out what he was referring to. Although we've won the race for outer space, Have the Soviets won the race for inner space? That's the question the CIA wants answered.

"They've got a humongous research program going involving the use of psychics, hypnosis, and psychology. They're being conducted not only here but in top universities, hospitals and research institutions around the country."

Greg repeated, "Where do I fit in?"

"Greg, I'm pulling you outta the EM lab and putting you on this contract." Ken leaned forward. "I want you to be my eyes and ears during tests Dr. Harold Puthoff and Russell Targ will be conducting here on Geller and other psychically talented people. I want you to review what the CIA has on what the Soviets have done and read everything you can find on the subject of astral projection."

"When do I start?"

"You just did. Geller's tests begin tomorrow morning at nine." Ken reached into his briefcase and pulled out a thick brown envelope. "Here's the CIA material. I'd like you to read through it tonight and return it to me in the morning. I'll be in early to tell Targ and Puthoff to expect you."

Greg set the envelope on the table. "Anything else?"

"Yes, I'll put you in touch with Jeff Southwell at the CIA on Thursday." Ken locked his briefcase. "Use your own common sense about how to proceed, and don't let on to any peers, except Puthoff and Targ, what you're about. I need results quickly, so let's meet here again two

weeks from tonight. I'll be anxious to see what you come up with."
Ken looked up at Molly approaching. "Do you want another beer,
Greg?"

"I don't think so. I'm anxious to get home and read this stuff. I'll
make a pit stop and then we can leave."

Greg maneuvered through the half dozen tables starting to fill with
people and walked past the bar to the men's room. He returned a few
minutes later and reached for his wind breaker and the envelope.

Ken said, "A guy at the bar just found you real interesting."

"Huh?"

"A tall well-dressed guy — looked like a company CEO type —
was staring at you like you'd been banging his wife." Ken pointed toward the door. "He hurried out as soon as you were in the men's room.
You got something going on you shouldn't?"

Greg raised his eyebrows. "Afraid not. I lead a pretty sheltered
life."

"I may be mistaken," Ken said, "but he looked like bad news."

Ken stood up. "Well, let's get outta here."

Ken looked around the parking lot and walked Greg to his car.

"As of now, Greg, you're part of the CIA Project Scanate." They
shook hands goodnight.

Ken stood by his several years-old Plymouth and watched Greg's
red MGA sports car pull out onto El Camino Real. Satisfied that no
one was following Greg, he started the Plymouth and headed for home.
He didn't notice a yellow Monte Carlo pulling into traffic two cars behind him. It followed at a distance and pulled up across the street with
lights out as Ken climbed the steps to his apartment door.

A large hand with manicured nails flicked a cigarette butt out the
driver's side open window as the car pulled away.

Chapter 7

After returning the CIA materials to Ken's office in the morning, Greg went to the physics lab. He entered an off-white meeting room informally furnished with a table and warm soft-leather chairs. Three men were already comfortably seated.

"Good morning, Greg." Russell Targ, about 30, a tall and lanky man with thick eye glasses stood and shook hands. "I believe you've met Dr. Harold Puthoff?"

A shorter man, a few years older dressed in light bell-bottom trousers, a turtle-neck sweater and dark blue sports coat, got up from the couch and said, "Please forego the doctor stuff, Russell. Greg's a doctor too, you know. Have you met our guest, Greg?"

"Yes. It's been my pleasure," Russell replied.

A tall slender man about the same age as Russell Targ stepped forward and slightly bowed european fashion. Greg shook a firm hand and looked into large brown eyes piercing out of a handsome face framed in thick dark hair. Uri Geller said, "I'm pleased to see you again, Dr. Philips."

Russell Targ spoke in a soft, soothing voice, "Uri, let's get started. This will be the program. We're going to move now to a 'Faraday Cage room' that will shield you visually, acoustically, and electrically from the outside world. We don't care what you do outside of that room, but while you're in it you have permission to be psychic."

Targ half-grinned and opened the door to the testing room. Before Geller entered, Targ reached in his jacket for a small tape recorder and switched it on.

"It is 11 o'clock, Wednesday, June 17, 1973. This is a remote viewing experiment with Uri Geller as the subject. Russell Targ and Harold

Puthoff as experimenters."

Targ switched off the machine. "Uri, once you're inside, one of our scientists you haven't met is going to open a dictionary in a room down the hall. He'll randomly select a word and draw a picture of what it has suggested. Your task will be to 'see' and draw on paper what our scientist has drawn."

Greg, Targ and Puthoff waited outside and looked in through a one-way window as Geller settled into a chair. The Faraday Cage room contained a single wooden chair, table, pad of blank paper, pencil, and a one-way intercom. Targ picked up a lab telephone, dialed two numbers, and said, "Uri's ready for you."

Several minutes passed and then Geller spoke into his intercom, "I see drops of water coming out of the picture ... I see purple circles ...." He paused, picked up the pencil and drew a bunch of circles, 24 in all. He set the pencil down and smiled.

Targ redialed and spoke into the phone. Two minutes later, a middle-aged portly man dressed in a white smock entered the room carrying a dictionary and a pad. He laid the open dictionary and pad before Targ and Puthoff. Greg looked over his shoulder. The dictionary was open to the word 'grape' and on the pad was his drawing of a bunch of 24 grapes. A collective sigh was heard from the three observers.

Two days later, Greg was back with the group outside the Faraday Cage room. Targ opened the informal meeting.

"Uri, you certainly impressed us all on Wednesday. So much so, we've designed a follow-up experiment to learn if you can 'see' over large distances as well. After you're in the test room, I'm going to telephone a colleague on the east coast who is awaiting my call — again with a dictionary. It will be the same program as last time, but now over 3,000 miles. Okay?"

Geller nodded once, sprang to his feet and entered the test room. As soon as Geller was out of sight, Targ placed a long-distance call to

New Jersey, told his medical doctor colleague to start the experiment and hung up the phone.

Several minutes later, Geller picked up the pencil and wrote the word medical. He paused, appearing dissatisfied and then wrote, organic ... living. He then drew a sketch.

"What the hell has he drawn?" Targ asked.

Puthoff focused through the window and shrugged. Greg turned and addressed the two of them, "I'm not an M.D., but the only thing that comes to my mind is an anatomical cross-section."

Targ walked to a desk and re-dialed the awaiting scientist in New Jersey. He listened attentively for several minutes and then sat down and murmured, "The word was brain. The drawing Bill just did in New Jersey was a cross-section of a human brain."

Greg said, "Geller's not a doctor. I bet he's never seen a brain in cross-section."

ELEVEN DAYS LATER AT THE RED COTTAGE

A rectangle of late sunlight was perceptibly lengthening on the Red Cottage lounge floor measuring out the afternoon.

"How are things going, Greg?" Ken raised an Anchor Steam in a mock toast.

"I'm totally wrapped up researching the subject of astral projection."

"How is Helen reacting to your long hours?"

"She hardly notices. What's going on in your life?"

"David called me from Berkeley this week." Ken's pale gaze drifted past Greg and scanned the patrons at the bar. "He's interested in working with me in some capacity when he graduates."

"You must be pleased."

"Very." Ken returned his gaze to Greg. "But, I'm anxious to hear what you dug up on astral projection."

Greg laid a well-worn briefcase on the table and rummaged through it. "Well, I told you about the Geller experiments."

Ken said, "I've already received Puthoff and Targ's report. Their report stated, 'Geller demonstrated his paranormal perceptual ability in a convincing manner."

Greg said, "The more I read about this subject, the more illusory it becomes." Greg pulled a small book out of the briefcase and handed it across the table. This book, *The Projection of the Astral Body*, is the 1929 ground-breaker on the subject. It says the astral body may be defined as the ethereal counterpart of the physical body."

Molly looked down over the bowl of pretzels she was about to deliver and asked, "Projection of the Astral Body, what's that about?"

Greg looked up. "Something that people like Aldous Huxley, D.H. Lawrence, Ernest Hemingway, and Charles Lindberg have all done."

Molly set down the pretzels. "And what's that?"

Greg had an impish grin. "Having a grand old time out of their bodies."

"Are you serious? Lindberg out of his body?"

Greg answered, "Lindberg said he existed out of his body at a point during his famous flight. He felt himself depart his body and pass through the plane's fuselage high over the Atlantic ocean. He was only connected to his body by a long-extended tenuous strand."

Molly said, "Are you pulling my leg?"

Greg answered, "I'd never do that, Molly. I've been doing a lot of reading on the subject. It's even reported in the Bible."

Ken's smile faded. "The Bible?"

Greg continued, "The Old Testament has the prophet, Elisha, voyaging out of body into the bedroom of a hostile Syrian king to eavesdrop on the king's military plans."

Molly made a clucking sound with her tongue. "It's never boring when you two come in." She walked back toward the bar.

Ken asked Greg, "Did Southwell have any CIA material that helped?"

Greg nodded. "The Russians have a guy heading up an astral projection espionage program named Leonard Vasiliev, a holder of the Lenin Prize. It's his research that's caused this small scale arms race."

"I've heard of him." Ken moved around the table to evade the late afternoon sun snaking through the blinds. He paused and whispered, "There he is again."

"Who?" Greg looked up at a frowning Ken staring at the door.

Ken settled into a chair. "The CEO type. I saw him looking at us when I got up."

Greg pivoted and half raised out of his seat. "Where?"

"He's gone now. Out the door."

Greg eased back into his seat. "Anyhow, I'm finding that astral projection and mind affecting matter may have common roots. There is even evidence that animals are involved. A new theory out of England called 'Morphogenese Fields' believes when some members of a group of organisms learn a new behavior pattern, it's instantly transmitted to all members of that species wherever they are. They've proven it by training rats in a lab in London to run a particular maze. Colleagues in New Zealand, with an identical maze, noted their rats overnight lost their confusion and manipulated their maze."

Ken said, "Stay focused on proving astral projection."

Greg nodded. "Next week we're testing a former Burbank, California police commissioner named Price who claims he used out-of-body ability throughout his career to capture bad guys. Hopefully I'll have some conclusions at our next meeting."

Ken reached for his wind breaker. "I'd better get outta here. Goodnight, Greg."

Ken drove home deep in thought about Greg's initial findings and didn't notice the empty yellow Monte Carlo parked across the street from his apartment. He put his briefcase under one arm while searching for his door key. He swung the door open into a darkened apartment and caught a scent of cologne as he slid his right hand along the wall for the light switch. A fist came swinging from the dark and slammed into his check. Ken dropped to the floor with a thud.

Several days later, Greg entered the Red Cottage lounge and saw Ken already seated in their favorite booth with two frothy beers and a bowl of pretzels in front of him.

"What in hell happened to you?" Greg looked at Ken's face with a bruised cheek and an eye almost swollen shut.

"When I got home the other night, I walked in on a jerk who cold-conked me." Ken answered.

Greg leaned forward for a closer inspection. "Why didn't you call me?"

Ken waved a hand. "I wasn't hurt."

"Did he take much?"

"It was strange. He ransacked the place and slashed open my briefcase, but he didn't take anything."

"Was it the guy you saw at the Red Cottage?"

Ken shook his head. "I don't know."

"Does it have something to do with our contract?" Greg slid slowly into the seat.

Ken sighed, "I don't know. I brought Southwell into it. His guys dusted and found excellent finger-prints that weren't mine."

"Then they'll find him," Greg nodded.

"Afraid not. The CIA, FBI, and local authorities don't have any records of the prints. It's like he doesn't exist."

"What does Southwell think?" Greg reached for his beer.

"He's put around-the-clock guards on my apartment. David's

coming down from Berkeley tonight to stay with me. He'll meet us here in a few minutes, so fill me in on the police commissioner before David arrives."

Greg took a quick sip. "Southwell suggested we test Price on the real thing. He gave us a longitude and latitude for Price to 'see'."

Ken rubbed his swollen eye as Greg continued. "Price gave accurate details of an underground satellite monitoring station in Virginia. Southwell's going to give him a longitude and latitude on a suspected Soviet missile site next."

"Southwell's been to see me." Ken interrupted. "The guys at the top of the agency are in contention on the whole idea of astral projection." He set his mug down. "Southwell's concerned this can backlash on us if we don't have some real science to explain it."

Greg sat facing a window. "David just pulled in."

"Don't mention the contract when David joins us, but give me your ideas on the subject as something we're both interested in."

Ken looked up at his son bouncing across the room and slid an index finger across pursed lips.

David stopped in mid stride. "Dad! Are you okay?"

David rested his hands on the table and studied his father's face from inches away. Ken pushed him away. "I'm alright."

David gently rubbed his father's cheek. "Who did it?"

"A chance encounter with a burglar." Ken jerked a thumb at Greg. "Last week I was worried my old buddy had an enemy - but not me."

David pulled up a chair, spun it around, and straddled it. "Hi, Dr. Philips. I'm honored to be included in your 'boys' night out'."

Ken looked up and raised out of his seat. "Molly, I want you to meet my son, David. David, this is our friend, Molly Schon."

David sprang up and shook her hand in both of his.

"My, you look just like your dad." Molly looked David up and down for a minute. "What can I get you?"

"A Coke would be terrific." David restraddled the chair.

Ken said, "Greg was about to tell me about a subject he's become interested in, David. You're welcome to listen in."

Greg hesitated. "I've been reading how parapsychologists are dealing with astral projection, a fancy way of saying being out of body, David. Some of them are tying psychokinetics, dematerialization, and astral projection experiences together as part of the same phenomenon."

David cocked his head. "You're interested in ouija board kind of stuff?"

Greg raised a hand. "These parapsychologists believe mind affecting matter during these phenomena may be a result of interaction at the primary level. They find parallels between quantum physics and psychic phenomenon."

Greg paused for effect. "Distance doesn't seem to be a factor in either. An event in one location can affect an event in another location miles away."

Ken said, "It brings to mind Heisenberg's Uncertainty Principle." David placed the palm of his right hand on top of the extended index finger of his left hand. "Time out. What's this Uncertainty Principle?"

Ken said, "There's a duality between particles and waves in which it's sometimes preferable to think of particles as waves."

"Do you mean particles like protons and electrons?"

"Yep." His father answered. "It was proven years ago in an experiment where a single electron was fired at a partition that had two parallel slits with a fluorescent screen behind it. Logic expected the single electron particle to pass through one slit or the other and be seen on the fluorescent screen. Instead, the electron passed through both slits, and was seen simultaneously at two positions on the screen. Obviously, the electron was a wave and not a particle during that experiment."

David said, "Way out. Does that mean the universe at the subatomic level isn't solid?"

Greg answered, "A protege of Einstein's named David Bohm has developed a theory that has no contradiction between physics and psychical data."

"How's that?" Ken asked.

"His theory doesn't separate mind from body. Rather, he sees a universal consciousness as the makeup of both. Things seeming to be separate in time and space are really linked together."

"How?" Ken asked.

"He sees the physical universe like a giant holograph with each part being in the whole and the whole being in each part."

David raised a hand. "I've heard about holographs, but, to be honest, I don't know what the heck they are."

Ken paused as Molly arrived with another round and then continued, "A holograph starts with a hologram, a two dimensional record of a 360 degree image of any object. It's constructed by the interfering waves of a single frequency wavelength beam of light that has bombarded the object. When a hologram is later illuminated by a similar beam, the object becomes visible in three dimensions."

David's face lit up. "Like the holographic dancers in Disneyland's Haunted Village."

"You've got it." Greg answered.

Ken asked slowly, "Greg, are you saying that our brain is a hologram interpreting a holographic universe? That what we see and hear around us is an illusion — not really there?"

Greg took off his glasses and rubbed his eyes. "This idea isn't something new, Ken. The famous philosopher, Bishop Berkeley, was already saying, *"All material objects, and space, and time are an illusion."*

He slid his glasses back up his nose. "It can explain how our minds have an instant effect across great distances ... and more importantly, out-of-body travel."

Ken shoved his beer away. "Drink up, everybody. David, would you wait in your car for a few minutes while I discuss something in private with Greg?"

Ken and Greg walked across the asphalt parking lot and stood under a tall light pole. Ken shoved his hands deep into his jacket pockets and looked at Greg with penetrating eyes. "Greg, we need to come up with a scientific explanation for the CIA on how a spy can be in two places at the same time. Is there a way you can validate your 'our brain is a hologram' theory?"

Greg answered, "I've got an idea, but it will take me some lab time to prepare an experiment proposal. I'm afraid it will require some complex computer modeling to interpret incoming frequencies from another dimension."

Ken looked over to his son patiently waiting in his Volkswagen. "I'm being pressed on this, Greg. Get me an experiment that can validate these hologram ideas or forget them."

Greg reached in his pocket for his car keys. "Meet me in my lab first thing Monday morning. I should have something for you."

SRI ON MONDAY

Greg got up an hour early on Monday and dressed casually in sports shirt and slacks. He wanted to warm-up the electron microscope and get it stabilized for the experiment he planned to show Ken.

He was finishing breakfast when Carrie came out of her bedroom. Tall, like her mother, Carrie was dressed in a pale-green pinpoint button-down blouse, denim skirt and carried a small calf-skin handbag.

"Morning, dad." She reached into her handbag and handed Greg an envelope. "My car insurance. Be a dear and take care of it?"

"Okay."

I'm off to work."

"Bye, honey."

Helen swept into the room in a faded cotton robe. Greg said, "I scrambled enough eggs for two if you'd like some."

Helen wrinkled her nose. "They look absolutely yucky."

Greg scraped them into the garbage disposal and said, "I'm working late again tonight."

"What time did you get home last night?"

"About 11."

She frowned. "I thought so. You must close your bedroom door gently when you come in late." She fixed him with a glare. "You know how lightly I sleep."

Greg poured his coffee down the sink and left for work. Once in the laboratory, he turned on the electron microscope and made a pot of coffee on the lab counter next to a hooded sink. He was sitting at the EM busily making adjustments when Ken barged in with David in tow. Ken said, "Greg, you mentioned during our last meeting that some computer modeling would be required. So, I've cleared David with Southwell and briefed him on what we're doing."

"That's great."

David said, "Ah ha, so last week's topic was more than just a subject you were interested in. Are we in the lab now to test your illusory world?"

Greg walked over to the sink. "Let's have some coffee and I'll tell you." He poured three cups and sat down at his gray metal desk.

"Taken together, Einstein's theory of relativity and quantum mechanics provide evidence that people do exist as conscious beings with free will. I think the universe is not a gigantic machine with us as helpless cogs, but something more akin to a gigantic thought. A sort of cosmic symphony that we're empowered to help conduct."

Greg focused on Ken. "I think it's God's universe and he allows us to participate with our brain holograms. He lets us help make them and then he triggers them into three dimensional holographs for us to see with his incoming wavelength frequency." Ken blew across the surface of his coffee. "Can you prove it?"

"If we can trigger a brain hologram with the same wavelength as God's incoming universal consciousness frequency, we should see a three dimensional holographic image of what the brain's owner was seeing the moment he died."

Greg sat his coffee down and walked over to the EM. "I think one of these might do the job."

Ken said, "All the holographic work I've seen has been done by laser beams."

Greg shook his head. "A laser beam's wavelength is far too big to see the tiny micro world these brain plates would exist in."
He patted the side of his electron microscope. "God's coherent wavelength must be very small. I think the wavelengths of electrons from an electron microscope are probably small enough. I think an EM can be tailored to do the job."

David raised his hand and waved. "How?"

Greg said, "Are you familiar with an electron microscope?"
"Nope"

Greg said to Ken, "Would you dim the lights?"

Greg sat down in front of the microscope with David and Ken looking over his shoulder. He pointed to the top of the tall nickel-plated cylindrical column. "Up here we have an electron gun. It has a filament similar to a light bulb's that emits electrons instead of light. The gun's electrons are 'shot down' through this column through a vacuum by a high voltage and focused into a small diameter beam by several electromagnet lenses. The beam then slams through the material it's going to magnify that is sitting on a specimen stage here."

Greg pointed toward a lever halfway down the column. He then turned to the control panel and pointed toward an indexed dial and meter. "This controls the gun voltage. The higher the voltage, the smaller will be the electron beam wavelength. The electrons are diffracted by the material they pass through and provide a magnified image of the material down here on a fluorescent screen." Greg slid his hand to point at a tilted window at the bottom of the column.

"I propose we make pseudo-brain hologram plates by cutting human brain tissue into micron thick thin-sections with a very sharp diamond knife microtome and put them into the specimen position of the microscope. We'll bombard them with a beam of electrons and look for any resulting three dimensional holographs from the brain's neural interference patterns at the fluorescent screen position — here." He pointed to the bottom of the column.

"We'll then go through the voltage range looking from wavelength to wavelength for any sign of an externally caused holograph."

David said, "Human brains? Do we get them at our local supermarket?"

"I think I can get them cut into thin-section plates and grid-mounted from a friend of mine, Dr. Donna Buckley, a neuro-pathologist at the V.A. Hospital in Palo Alto. She's received permission from a number of families to use nervous tissue — brains and nerves — of their departed ones to further science." He stopped and turned to Ken. "I think an electron microscope might provide the science we need to give the CIA a glimmer of our holographic universe."

Ken smiled doubtfully and said, "And that in turn may give some answers we've been searching for from our earliest days, first expressed in mankind's interest in Astrology, Reincarnation and Dreams?"

Greg sat back and spread his arms. "As Saint Francis of Assisi put it, 'What we are looking for is what is looking.' "

The next morning, Greg knocked and entered the Veteran's Hospital's microtome room. Dr. Donna Buckley greeted him. "Hi, handsome, here for your favor?" She was a little older than Greg's daughter with a dimpled chin and thick chestnut hair.

Greg said, "You know, Donna, it's hard to imagine what you must do to a cadaver to end up with mounted brain tissue."

"It's a rotten job but someone has to do it. Do you want tissue from a young or old vet?"

"Just that he was lucid at the time he died."

"I can have them for you in a week," Donna said as she gestured toward the microtome. "Promise to share your Nobel Prize with me."

Greg laughed.

Chapter 8

RIVERSIDE - THE THIRD DAY (Wednesday)

reg picked up his glasses in the late afternoon and drove to the football field. Emily was waiting, this time dressed in a blue pleated skirt and white blouse buttoned to her neck.

"You look quite distinguished in those glasses." She ducked into the car, slammed the door and turned to Greg with a radiant smile. "Maybe a bit rakish," she said.

Greg looked her over and then pulled away from the curb. "You look lovely."

Emily smiled. "Thanks. Now, I want to hear why you're here."

Greg parked the Chevrolet on the street and walked with Emily through the old inn's entry path covered in large adobe brick and lined with 18th century cannons. They entered a courtyard dining area and were seated by an elderly waiter dressed in a faded but starched shirt and black trousers, shiny in the rear.

They ordered. Greg took a sip of water and gazed at Emily over the glass. Her hair this day was swept to one side of her head, adding a touch of mischievousness to a face that otherwise had the innocence of a child.

"I'm here because of a government contract we had at Stanford Research Institute. Ken Hoard was the laboratory manager and my direct supervisor for the first few years."

"Tell me about it."

"We were to use an electron microscope to find a scientific explanation for a new type of espionage the government was interested in."

"Did you?"

"We were accused of botching our results. Ken was fired and I was sent back to the lab in scientific disgrace," Greg explained. "It didn't end there, however. Our government contract manager, a guy named Southwell, told Ken his dismissal was sophistry."

Greg paused while the waiter served them. Greg picked up a fork and waited until the waiter moved away. "The Director of the agency wanted Ken to pursue the experiments in a more secure environment than we'd had at SRI. And because of the outrageous nature of our results, he was concerned about continuing to use congressional funds."

Greg nodded at Emily's plate. "Let's eat before it gets cold."

"It is cold. Chicken salad is always cold," Emily joked.

Greg grinned, took a bite and wiped his mouth with a napkin. "The Director had personally contacted a very patriotic and very wealthy Texan who just happened to be a good friend of the current administration. This guy agreed to independently fund Ken's continuing research. He set him up in an innocuous old house in the hills of La Honda that looked like a weathered redwood barn from the road. The Texan gave him a generous budget to continue the research. After Ken's son, David, graduated, he joined his dad."

They finished their salads and the waiter seemed to materialize. "Can I get you some coffee?" the waiter asked.

Greg looked at Emily and raised his eyebrows.

"I don't believe so, thank you."

Greg pushed back. "I'm going to the men's room. Could I interest you in a rowboat ride around Evan's lake in Fairmont Park?"

"That sounds like fun, but first I'd like you to meet someone."

Greg frowned and excused himself. As he strode past the row of sinks, he saw both urinals on the far wall were occupied. He hesitated, and waited for a skinny young man and a broad-shouldered adult to finish. The young man zipped up and turned toward the sinks. Greg froze. It was his living image from another time! The broad-shouldered man turned and scrutinized Greg carefully with a predator's eye. It was the stranger from the De Anza Hotel lobby.

Greg spun around and fled the room. He returned to the table and quickly gathered up his bill, perfunctorily helped Emily into her coat and hustled her toward the cash register.

Emily tugged at his jacket. "Did you see my student? I saw him go into the men's room a minute before you."

She gave him a concerned look. "Just a minute. You look like you're about to have a stroke. Was it the boy? I asked him to join us here."

Greg used the time, rapidly walking to his Chevrolet rental, to compose himself. As he pulled out into traffic he turned to her and answered, "Yeah, it shocked me. He really does look like me ... but he didn't seem to notice ... I don't know what to say."

Emily studied him. "I really wanted you to meet him."

"Well, I just did. And he doesn't know me."

Emily held her breath with her eyes locked on his.

Greg smiled ingeniously. "How are you on time?"

Emily exhaled. "I'm free for the rest of the day."

He pulled into a liquor store at Third and Main. "Then let's do this right."

He came out with a bag and drove to the De Anza hotel.

"Sit tight for a minute. It can be chilly on a lake this time of year. I'm going to borrow a blanket off my bed. Be right back."

Inside his room, he started to strip off a blanket when he noticed his canvas suitcase was open and the newly purchased clothing was strewn across the bureau. Assuming a nosy maid, he started refolding the clothing when he noticed the ash tray on the bed stand. A single Old Gold cigarette butt was stubbed out and left in the tray. His adrenalin flowed. "Who is this guy," he whispered as he began cramming his belongings back into the suitcase. He grabbed it up, and dropped his key and four twenty dollar bills onto the bed.

At the car, he quickly threw the suitcase and blanket into the trunk, and drove the few blocks to the lake. On arrival, he removed the blanket before opening Emily's door. A few minutes later, Greg looked about apprehensively and then pushed a rented rowboat, with Emily sitting in the bow, away from a small loading dock. He rowed toward a palm-covered island in the center of the lake.

"Tell me about yourself, Emily."

"I'm an only child. My parents are dead."

"I'm sorry. Do you have other relatives?"

"An aunt and uncle in Illinois whom I've only met once."

"How come you've never married, Emily?" He noticed there were only a few other boats on the lake.

She shrugged her shoulders. "Fussy, I guess."

Greg steered the boat into the grassy slope of the island and stepped out. They selected a wild, grass-covered area in the middle of several palms that provided modest cover from the shore and spread the blanket and emptied the contents of the shopping bag. He'd purchased a chilled bottle of champagne, glasses, napkins, knife, crackers, and cheddar cheese.

Emily tucked her skirt under her knees and knelt to open the cheese. "What became of Ken's research?"

"Ken and David outfitted a world-class scientific laboratory, including a new electron microscope but had no success. Ken told the Texan his only hope would be an electron microscopist expert. The Texan told Ken to bring me on. I was to stay until I proved our theory one way or the other. He told Ken money wasn't an issue, so I was offered double my SRI salary to take a two-year sabbatical. I leaped at the money and the opportunity to work with Ken again."

"Were you successful where Ken wasn't?" Emily unwrapped the

cheese and started cutting it into small cubes.

Greg opened the crackers and made miniature cracker sandwiches with the cheese. "Yes and no. I found something absolutely amazing, but it wasn't what I was looking for"

"And that was?"

Greg hesitated indecisively and then answered, "A miniature black hole."

"A black hole?" Emily jerked up her head. "I've never heard of such a thing."

"No. I knew you hadn't." He bit into a cracker sandwich. "There's nothing more I can tell you."

Emily set her empty glass down.

"What happened to your marriage?"

Greg paused and thought about his answer. "I met Helen while I was stationed at an airbase in Savannah when I was 19. She was fun to be with. Had a very well-to-do, deep-south family. I was lonely and we got along famously. But once we married, and I was in college on the G.I. Bill, things weren't so great." He paused and added, "Moving from a columned home in Savannah to a studio apartment in Pasadena was a bit of a shock for her." Greg paused again. "To be fair, I'm sure we both contributed to our problems."

"Do you still love her?"

Greg said, "No ... maybe the girl she once was ... but, no."

Emily said, "Why did you stay with her so many years?"

"I got her pregnant while I was in grad school and decided to stay until our daughter, Carrie, was raised."

Emily asked, "She's raised now, and you've definitely left Helen?"

"I'll never see her again." Greg sat cross-legged and reached across the blanket and gently pulled Emily to him. Emily laid her head on his shoulder.

The sun was filtered by the palm branches and the lake gleamed in the last sprinkling of light across the water.

Greg turned Emily's face toward him. Their kiss was gentle. They pulled away and looked into each other's eyes. Greg kissed her eyes and ears and neck. Emily moaned softly.

Greg lowered her partly onto the blanket and partly into the tall grass. Emily leaned back and said, "No, not here," and pushed him away. She regained her composure, stood and brushed down her dress. "I can't risk being seen like this. Come home with me."

Chapter 9

LA HONDA - SEPTEMBER, 1981

G reg sat at a checkered, linoleum-covered wooden table across from Ken and David in the rustic kitchen of their research facility. The walls were covered in weather-stained knotty pine paneling, and the coffee they were sharing had been brewed on a white 1940s gas stove.

"It's heating up again, Greg. The chairman of the House subcommittee on intelligence evaluation and oversight recently went on record in Congress, saying the Soviet secret missile site our Burbank Commissioner saw remotely has been confirmed by aerial photography. Congress's interest is peaked again and Tex is pushing me."

Greg stood and thoughtfully walked across the room to a window overlooking the redwood trees. He turned and said, "I've been working with a new software program I asked David to prepare. I think I've found our problem in generating a holographic image."

Ken said, "What is it?"

"I think we need even shorter electron frequency waves to sample. Our EM peaks out at 100 kilo volts and has only five fixed voltages between 20 and 100. I feel fairly confident the correct wavelength will be found at a very precise voltage higher than 100KV, maybe as high as 220 KV. Our justification to the CIA for the astral projection tests was predicated on the brain's deep structure being hologram plates we could pass electrons through and generate a three dimensional image of our reality. But, the electrons had to have the same wavelength as the incoming universal consciousness wavelength that had formed the image."

"So what do we do?" Ken asked.

Greg said, "To be sure we don't miss the wavelength, We need an EM with a range of 20,000 to 220,000 electron volt imaging capability in one-volt increments. With that flexibility, we can search through 200,000 individual wavelengths."

Ken shook his head. "No such microscope is commercially available."

Greg said, "But the technology exists to build it."

"Can you build it?"

"Not a chance. It will require an EM factory with the finest electromagnetic lenses and access to the most stable electronic components. But I could prepare the technical performance specifications. We're going to have to vary a lot of controls thousands of times for the tests, so computer control is a must. That's where David will come in, specifying the degree of automation required."

Ken drummed his fingers impatiently on the table. "What will it cost?"

"Assuming I can find a company to build it, and also willing to amortize part of the development costs over future sales, maybe a million dollars."

Ken winced. "Ouch. Well, let me talk it over with Tex. You two go ahead and get started on the specs."

A MONTH LATER

Greg stretched his legs and tried to get comfortable in the tight confines of the JAL 747 business class seat. He looked at his digital watch and concluded they were about to cross the international dateline. They'd been in the air just over eight hours with five more to go to their destination.

He thought back at his flurry of activity with David after his critique of their failed holographic tests. First, Ken had gotten approval from Tex to spend up to one million dollars on the special microscope he and David had specified in detail. Next, he'd visited the U.S. Sales Office of Hitachi Ltd., the biggest high voltage electron microscope company in the world. Its general manager, Dr. Robert Buchanan, was his old friend.

Greg recalled their conversation in Buchanan's office that resulted in this flight. "Bob, you've got to convince your factory this is a project worth doing. Once it's completed there will be a multitude of biological and metallurgical applications for a microscope with this versatility. And if my holographic tests are successful, a huge new market will open up."

Greg and Buchanan sat across from one another over a highly polished tea table in Buchanan's office. A large Japanese watercolor screen of Mount Fuji decorated the facing wall.

Buchanan said in a mild brogue, "I understand what you want, Greg, but I don't have the authority to approve it, and I know the factory in Mito City isn't into making markets for new tools. I recommend you go to Japan and sell your idea directly to the factory manager, Dr. Shoichiro Shinohara. I get along well with him and will write in advance and introduce you. That's all I can do."

Later, Buchanan called with open visit dates and told Greg a translator named Ned Shikashio would meet his flight at Tokyo's Narita International Airport and escort him to the Naka Kojo factory in Mito City.

A stewardess interrupted Greg's reverie and offered him a pillow. He slept for the rest of the flight.

Twenty-four hours later, exhausted from jet-lag and dejected from a day at the Naka Kojo EM factory, Greg sat with his translator in the small lobby of the Mimatsu Japanese Inn.

Ned Shikashio poured hot sake from a small decanter into miniature glazed sake cups and handed one to Greg. "I'm sorry, Greg. Shinohara-san can be stubborn, and it's just not part of our culture to develop innovative products without a proven market." Ned raised his sake cup. "Kompai."

Greg nodded and looked out the window of the Mimatsu Inn at a manicured Japanese garden. A limpid stream flowed gracefully with syllables of water on stone. Miniature bonsai trees and chrysanthemums were aesthetically placed around the garden. The Mimatsu was a traditional inn, as Greg had requested, and he had slept poorly the previous night — on a futon on top of a thin tatami mat on the floor.

"Can I make an overseas call?" Greg asked Shikashio. "I've got to report the bad news to my boss."

"I'll talk to the Mama-san and have her set it up."

"Yeah, I can hear you just fine." Ken said. "Don't give up yet, 'ol buddy. Our Texan has pull at the highest levels of our government. I'll call him and see what he can do. Give me your telephone number and stay where you are until you hear back from me."

The Mimatsu Mama-san awakened Ken at six the next morning with an incoming telephone call. Greg pulled on the room yukata, a light blue cotton robe, slid his toes into the hotel slippers, two sizes too small, and shuffled after her down the hall.

"Aragato," he thanked her.

"Greg, Tex just called. He's been on the horn to Washington all day and struck pay dirt. A very high government official has talked with his counterpart in Tokyo. Your factory manager just changed his mind. You'll have your microscope in nine months."

LA HONDA: NINE MONTHS LATER

Greg stood behind a Japanese engineer in the converted high-ceiling living room of Ken's research facility. It was nearly filled with a nickel-plated electron microscope gleaming with orange, red, and green miniature bulbs in the dimmed lighting. The engineer said, "Dr. Philips, I'm ready to show you our electron microscope meets your specifications."

Two days later, Greg signed off on an acceptance test. Tex would now process payment on an international letter of credit. David arrived and Greg said, "Let's go to work. You're going to have to provide a great deal of automated computer control in your software program, because we don't know what the microscope conditions will be to trigger a 3D image; we don't know what wavelength we're looking for, at what magnification we're likely to see the image, or the required density of bombarding electrons we'll need to get us enough signal above noise to see it if it's there."

David said, "You'll want the computer to control all of these variables at each wavelength while you look for a holograph to appear below in the vicinity of the fluorescent screen."

Greg agreed. "I also think we'll better our chances if we look for hologram plates in four pi-steradians in different orientations of the brain."

David said, "You mean like cutting a grapefruit in cross-sections as it's rotated?"

"That's right. I'll give Donna Buckley a call."

Chapter 10

RIVERSIDE - THE FOURTH DAY (Thursday)

reg awakened in an unfamiliar double bed. A ray of morning light had meandered through a crack in the Venetian blinds and spread across his face. He looked about at a neat feminine bedroom. The walls were painted in a warm shade of light blue, and the partially pulled Venetian blinds were covered in antique white diaphanous curtains. The morning was dark with a charcoal gloom. He swung his legs to the floor and looked for Emily. He found a note for him taped to the bureau mirror.

Good morning. You were sleeping so soundly I didn't want to wake you. I'm off to school, will be back after three this afternoon. I left the coffee pot on, juice and sweet rolls are in the fridge. A spare house key is on the kitchen table.
Love, Emily

Greg dressed, cracked the front door and peered out onto the porch. No neighbors in sight. He walked out to his car and returned with his suitcase, a tool kit and the items he had purchased from the wrecking yards on Tuesday. After a shower and shave, he spent until mid-afternoon carefully disassembling and disabling the items. He was pleased with the results. Even a close inspection would not reveal his tampering.

He peeked through the blinds at the dead wintry day below, and with the coast clear, locked up and walked to his car. His second task of the day was to return to Hogles Brokerage.

"My, but you're having a nice run of luck, Mr. Philips. Both Gaspe Oil Ventures and Nevada Tungsten have doubled in value since you purchased them on Tuesday," the floor broker commented.

"That's great. Now I'd like to buy $7,500 worth of Trad Television."

"Another penny stock. I'm tempted to follow your lead and buy some myself!"

"Tomorrow is the fifth day since I signed up, so I will have satisfied the exchange statute and can collect some money. After buying Trad, my account should be a little over $3,000. I'll be by before lunch and would appreciate a check for $3,000."

On the way to Emily's, he took a pass through the hamlet of Highgrove, recognized a familiar access road, and a quarter of a mile into the groves spotted what he was looking for: a Sunkist Company loading dock and parking area. He turned around and returned to Riverside. Emily was waiting for him just behind her door when he arrived home.

"Where did you put your car?"

"I thought it best to park it in your garage away from your neighbor's prying eyes."

"Good! I just had a bad experience as I was leaving the school. A man stopped me on the street and demanded to know your whereabouts."

"My whereabouts? What did he look like?"

"Big, early middle age. No friend of yours."

"What did you—"

"I told him I hardly knew you. He said he'd seen us having lunch yesterday."

"He must have been at the Mission Inn. There was a big guy in the men's room when I saw your student."

"I told him we'd met by chance a couple of times. That you were thinking of moving here. You wanted some advice and were gracious enough to buy me lunch in return for that advice."

"Did he believe you?"

Emily shivered, "I don't know. He said you were a dangerous man that he must find. He gave me his name and room number at the De Anza Hotel and ordered me to contact him the minute you show up."

"What's his name?"

"Herman Hoyer. He said he hadn't been sure you were the man he was looking for until you ran from him at the restaurant. By the time he looked for you at the hotel, you'd left without checking out."

"I don't know any Herman Hoyer. It has to be a case of mistaken identity. He must have seen my reaction to seeing your student and thought I was reacting to him."

Greg carefully selected his words. "Remember, I told you my business here is to make amends for something I did long ago. That'll happen in three days. I've got to avoid anybody getting in my way between now and Sunday. I'll tell you why then."

"He made it clear he's going to hurt me next time if I haven't co-operated."

Greg said, "Come with me, and there won't be a next time."

Emily shivered again. "I'm not going to let you out of my sight."

"We'll wait until after dark. He probably knows my car, so I'll trade it in at Hertz for a different model."

"Where will we stay?"

"We'll go to a hotel in Colton. You call in sick tomorrow morning. Pack up clothes for four days."

Greg periodically cracked a lower corner in a Venetian blind and scanned the street. A little after five, he saw a black sedan pull under a shedding eucalyptus tree across the street. He watched the car for an hour but nobody got out. At 15-minute intervals, a match would flare as a cigarette was being lit. Emily's house was being watched. They slipped out the back and two hours later were relaxed in a rear booth of a restaurant in neighboring Colton. They had exchanged the Chevrolet for a generic Plymouth at Hertz and checked into the Colton Inn under aliases.

"Phil, you know I'm taking a lot on faith. Please don't let me down. I'll expect a full explanation on Monday."
Emily seemed to be reevaluating her position.

"I really wish I knew what this guy is after. I don't have any idea who he is. It's got to be a case of mistaken identity."

Emily sat back. "This is pretty exciting stuff for an old maid school teacher."

No further speculation was in order. They simply didn't have enough information. Over dinner they fell into small talk.

Emily said, "Did you know that William Holden is in town? He's shooting a new horse racing film at the old De Anza track north of Fairmont Park. It's going to be called Boots Malone."

"No, I didn't."

"Did you hear the flying saucer rumor?"

Greg stiffened. "What are you talking about?"

"Several people apparently called the Highway Patrol Tuesday evening and reported seeing something that looked like a flying saucer hovering in the distance over the Santa Ana riverbed several miles north of town. The Highway Patrol checked it out and didn't find a saucer, but did find some peculiar tracks in the dry river bed."

"Is that all?"

"Yes, I think it's probably just a prank."

Chapter 11

LA HONDA: AUGUST, 1982

G reg charged into Ken's office. "Ken, got a minute? I want to show you something on the new microscope."

"What is it? Do you have a holograph?" Ken yanked his boots off the desk and dropped the paper he was reading.

"I don't know what I have. It's in a brain tissue image on the fluorescent screen. Something I've never seen in an EM. Something that shouldn't be possible."

Ken sat at the instrument in the darkened EM room and looked through the lead glass viewing window at the EM image displayed on the fluorescent screen. He saw a typical highly-magnified image of brain tissue just as they had been seeing throughout the last three months of searching for holographs. Except this image was obscured in the center one third with an apparent dynamic maelstrom of activity. The brain cells were swirling and spiraling and accelerating into a nadir point at the center of the screen.

Ken pushed back from the instrument. "It looks like a turbid whirlpool. What in hell's going on?"

"I don't have the foggiest. I've never seen anything like it.
EM images are always static. It started happening this morning when I made the electron gun adjustment from 199 KV to 200 KV. All the images I've seen for the last three months have been what I would expect from brain tissue. Just the contrast difference you'd expect as I've crept the voltage up from 20,000 volts one volt at a time. Now watch this."

Greg backed off the voltage one volt. The dynamic maelstrom effect came to a stop. He then moved the voltage back to 200,000 volts and decreased and then increased the magnification. The maelstrom stopped, started, and stopped again.

"So it only occurs at a very precise voltage, magnification and beam current," Greg noted.

"Well, you sure as hell have found something. Do you think you've hit on the right wavelength? Could it be an aberration of a hologram's holograph?"

"I'd be kidding you if I said I had any idea about what's going on," Greg responded.

"I think you'd better find out what's going on in there before going any further."

"I agree," Greg said. "I'll call around to my EM friends and see if any of them have ever observed this phenomena."

Greg spent the next week in the Stanford University Engineering Library reviewing every EM paper published since the Electron Microscopy Society of America's (EMSA's) first meeting in Chicago in 1942. He telephoned the movers and shakers in the field, people like Fabian Pease of Stanford and David Joy of the Bell Labs in New Jersey. The polite consensus was his EM was probably malfunctioning.

Disappointed, he took the next pass through all biological EM work by all American scientists during the same time frame. After that, he turned to work reported in other countries. He hit possible pay dirt in a paper written by a material scientist from the University of Toronto in 1947 and another from a Tohoku University biologist in 1980. Elated about his Canadian find, Greg returned to La Honda and reported to Ken.

"A material scientist up in Canada may have seen what we have over 30 years ago. Dr. Paul Pratt, of the Material Science Department of the University of Toronto. He apologized in the text of his paper for the poor quality of his micrograph. He was trying to study monocrystaline nickel films in a homemade EM of his own design. Remember, there were very few commercial systems around then."

"What did he see?" Ken asked.

"His paper centered on his difficulty imaging the films due to lack of penetrating power in his EM. So he jerry-rigged a 200 KV power supply to his gun and was able to image for a few minutes before the gun started to arc over in the poor vacuum system they had in those days. I made this copy of the micrograph he apologized for. Here, does that look familiar?"

"Pretty fuzzy, but it could be what we're seeing."

"I'm going to try to locate Pratt."

"Probably a long shot after all these years; anything else?"

"Maybe a Japanese biologist, but I'm not sure. His paper didn't include a micrograph, but he does note in his closing paragraph a strange dynamic phenomenon while imaging cell mitochondria at precisely 200 KV on a JEOL model JEM-200.
He concluded he would study the phenomenon further and report at the next meeting of the society. This was just two years ago."

"Did he?"

"No."

"The common denominator seems to be the gun voltage. You're obviously on the right track," Ken observed.

In the morning Greg placed a call to Canada. "Hello, Dr. Harloe. I'm Dr. Greg Philips, recently from Lawrence Radiation Laboratory in California."

"Yes, what can I do for you, Dr. Philips?"

"I understand you're the chairman of the Material Science Department there in Toronto. I'm trying to locate a professor who was a member of your faculty 30 years ago. Do you have records that may give me a place to start looking?"

"What was his name?"

"Paul Pratt. He apparently built his own electron microscope at the time and published some results. I'd love to talk with him about it, if he's still around."

A pregnant pause came over the wire. "I'm afraid you're out of luck, Dr. Philips. Paul disappeared nearly 30 years ago without a trace. I know, because I was an undergraduate here at the time and took one of his classes."

"Disappeared without a trace? Was it foul play?"

"It caused quite a stir at the time and made national headlines for a few days. As I recall, he was very excited about something that had come up in his work. He started neglecting his classes and locked himself up in his EM laboratory. He became very secretive, and we students started speculating about his stability. Then one day he was simply gone."

"When did this all take place?"

"Right after he went down to the States and gave a paper at the Franklin Institute in Philadelphia."

"Right after he presented the nickel film results of his 200 KV micrograph," Greg whispered.

"Was he married?"

"Single, but married to his microscope. I recall a member of the faculty working late met him outside his lab the night he turned up missing. He said Paul was very excited about installing a milling accessory he'd made onto the microscope. Paul was quite a gadgeteer. He was going to use it to mill away layers of material from his metallurgical samples while the microscope was imaging them. Cutting edge science in 1947; old hat today. He was gone in the morning and hasn't been seen or heard from since."

"Were his belongings gone from his home?"

"No, it was the strangest thing. Nothing was disturbed in his apart-

ment. Oh, and the sample chamber and stage of his electron microscope were gone, as well. The mounting screws had been sheared. It
would have taken an enormous force to yank them out that way. We
speculated Paul had rigged up a chain hoist to accomplish it."

"Why would he have done that?"

"I'm afraid he was quite mentally ill."

"What caused you to conclude that?"

Dr. Harloe laughed. "I'll give you an example. He had adapted a
very high-voltage supply to his homemade microscope. Paul confided
to a friend that whenever he ran the microscope at the highest voltage,
he felt irresistibly attracted to the microscope." Harloe laughed again.
"How's that for being married to your microscope?"

Greg said slowly, "Could I see it? Does the university still have
it?"

"Oh no. It was cannibalized years ago by grad students for other
projects. May I ask for your sudden interest after all these years?"

"I've been observing some strange EM images and did a literature
search. I think Dr. Pratt observed the same thing 30 years ago shortly
before his disappearance."

"That's interesting. I had a Japanese biologist call a couple of years
ago and ask the same questions. He was particularly interested in the
ion beam milling accessory. I don't know why. It would have no ap-
plicability on biological samples."

Greg scurried into Ken's office and explained the peculiar disappearance of Dr. Pratt.

"I wouldn't place too much emphasis on his disappearance. It's
not unusual for guys to bail out like that. He sounds a little strange.
You'd better move on to the Japanese biologist."

Greg tried unsuccessfully for three days to telephone Dr. Sus
Honjo. Meanwhile, he experimented with a variety of sample types in
the EM. The material of the samples made no difference. A maelstrom

suddenly formed on the fluorescent screen whenever the parameters were satisfied for voltage, electron density, and magnification.

He removed all of the samples and placed them under a lower power light microscope sequentially. Each sample was missing a circle of material where the electron beam had bombarded it. He called Ken in again for consultation.

"It's getting stranger all the time, Greg. I think it's imperative you get hold of this Dr. Honjo. He's had two years to deal with the phenomenon, if he saw the same thing."

"He hasn't returned my calls, but we're having the annual EMSA meeting down at Asilomar in Monterey in two weeks, and he's one of the invited speakers. I'll track him down there. I've prepared a 200 KV electron micrograph of the same type of mitochondria tissue he reported on two years ago. My micrograph shows the same maelstrom effect I think he observed. I'll show it to him and find out if we're talking about the same thing."

Two weeks later, Greg told Helen and Carrie over dinner he would be gone for three days to the annual EMSA conference. Helen said, "Gregory, before you go, take care of the washing machine. The spin cycle won't start automatically. I have to jiggle the knob to get it to work."

In the morning, Greg left before his wife and daughter were awake. Only his dog, Pee Bee, hobbled awake and wagged his tail goodbye. Pee Bee was 15 and coming to the end of a happy life as Greg's faithful companion. Greg had named him after a druggist character in a 1940s radio show from his childhood called The Great Gildersleeve.

Greg cruised his MGA down Highway 1 to Monterey. He was looking forward to three days on the beautiful Monterey Peninsula. He needed to get away. Greg had devoted the last two weeks taking the brain tissue through the last 20,000 volts between 200 KV and 220 KV without a trace of a holograph and was dispirited. He needed to learn if there was any tie between the maelstrom at 200 KV and the absence of a holograph.

After Santa Cruz, the highway veered west toward the coast through lush fields of artichokes and broccoli. Greg pulled into a road-side artichoke stand and feasted on a paper plate full of steaming hot, deep-fried artichoke hearts. Thirty minutes later, he pulled into Asilomar and slowly cruised through the grounds. He observed two-story buildings with names like Windward, Afterglow, Nautilus and White-caps. Their weathered redwood siding blended with the timeless beauty of a wind-sculptured forest of ancient Monterey Pines surrounding them.

Asilomar is a state park that caters to conferences and vacationers. Greg checked in and asked if Dr. Sus Honjo had arrived. He was told Honjo was scheduled for a late arrival. Greg was given an EMSA schedule and saw that Honjo's talk would be at 2:30 the following afternoon.

He met Donna Buckley in the food line at breakfast. Greg and Donna sat in on several biological papers while he waited for Honjo to appear. At 2:30 in the afternoon, he saw an elderly Japanese gentleman, with the aid of a cane, climb awkwardly up the steps to the podium. Honjo was a short, soft and doughy little man. He presented a boring paper in broken English describing the properties of cell cytoplasm. There were no questions at the conclusion of his talk, and the session took a break.

"Excuse me, Dr. Honjo. Could I have a moment to talk with you?"

Honjo asked, "What can I do for you?"

"My name is Dr. Greg Philips. I'm very interested in your work. Could we go outside and talk?"

"Yes, Dr. Philips. Are you studying cell cytoplasm?"

They settled on a bench under a Monterey Pine. A cool light breeze murmured through the surrounding shrubbery. "No, Dr. Honjo. I've been looking at human brain tissue in my EM at voltages ranging from 20 KV through 220 KV. At 200 KV, I am observing a phenomenon I don't understand."

Honjo sat up, his knuckles were white holding his cane.

Greg continued, "In a literature search, I came across your results." Honjo struggled to his feet, "I cannot help you."

"Wait a minute Doctor, is this what you saw?" Greg held out his micrograph.

"Get away from me. Leave me alone!"

On Monday, David followed Greg into Ken's office, the converted master bedroom with yellow flowered wallpaper.

Greg said, "Honjo was horrified the moment he saw my micrograph. I got nothing out of him."

Ken motioned for the two of them to sit. Greg continued, "But I've been doing some thinking and reading about what we're seeing going on in the EM."

Ken said, "You mean the whirlpool with material missing at the center?"

Greg nodded, "Do you realize how similar this phenomenon is on a small scale to what astronomers think black holes in space are like."

"Black holes?" David blinked.

"Are you familiar with the term?" Ken asked his son.

"Sure. Since the Disney movie last year, everybody is. They happen when an aging star blows itself apart as a super nova while its core remains whole. Gravity gets so great that it pulls everything into it, including light."

Ken said, "Are you saying you've made a miniature black hole?" Greg smiled, "Hamlet said, *'There are more things in Heaven and*

Earth, Horatio, than are dreamt in your philosophy.'

Ken turned to Greg. "I can understand how a huge neutron star collapsing under its own gravity can become a black hole by curving its light back onto itself, and I can understand that the crushed star can be incredibly small with gravity so strong it will pull anything into it that comes too close. But a miniature black hole?"

"Read this." Greg took a small reprint from his briefcase. "Stephen Hawking, of Cambridge University, predicted 10 years ago that the universe is populated with submicroscopic black holes." He handed it to Ken.

Ken set his coffee cup aside and scanned the paper. "Do you think the whirlpool we're seeing is coming from within an event horizon of a miniature black hole?"

David made a timeout sign with his hands. "What's an event horizon?"

Ken answered, "An imaginary envelope around the collapsed star where everything outside the envelope is safe, but everything inside is doomed to be drawn in and squashed into the center by the tremendous gravitational pull."

Greg said, "Now doesn't that sound like precisely what we're seeing?"

"If it is," Ken answered, "you've made the discovery of the century: a man-made miniature black hole."

"Ken, you know mass is not the only source of gravity. Pressure, stress, and energy also contribute. I think I've made a mini-black hole by directing an extremely intense beam of very high energy-pressure electrons into a tiny pellet of matter. They've somehow provided the right combination of pressure and energy from the electrons high voltage and stress from their highly focused condition, to form a mini black hole."

Ken paused and thought through his next question. "How do you propose to prove this theory?"

"By looking for an event horizon," Greg said. "If we do have a miniature black hole, then it's going to have a miniature event horizon surrounding it. I'm going to add a micromanipulator to the sample stage in the EM that will allow us to move a second bit of matter slowly toward the whirlpool while we're looking at the EM image. If I'm right, as we push the material through the event horizon, it should be drawn the rest of the way to the center of the whirlpool and disappear as it is squashed down infinitely small."

"Where do you think our holograph is?" Ken asked.

Greg sat back. "Maybe we'll find it in the process of understanding if I've made a mini black hole or not."

Chapter 12

RIVERSIDE - THE FIFTH DAY (Friday)

G reg and Emily awakened Friday after a restless sleep. It had started raining during the night, and Emily had repeatedly bolted awake and peeked through the drapes for Herman Hoyer.

The arrival of Hoyer and his accusations against Greg were haunting her. Tired and irritable, she accompanied Greg to Hogles. Silence permeated the Plymouth. The only sound was from her raincoat making a crinkling noise as she squirmed over the fabric-covered seat.

Greg flicked on the wipers and looked at her out of the corner of his eye. "Why so quiet?"

Emily's red-rimmed eyes stayed fixed on the wipers. "I don't feel like talking."

Greg turned his head to study her fully for a moment and then turned on the radio. The car filled with a disc jockey's theme song: *"Hey all you cats on the way to school, better play it hot, better play it cool. Keep your minds on the books like the teachers do. Let Dick Whitinghill watch the clock for you. KNBC Los Angeles."*

Greg pulled into a parking lot and turned off the engine. "Look, I've got to go in here to make a trade and get a check. It will take me upwards of 30 minutes. I don't see any sign of Hoyer. Why don't you run into Mape's and get a bite to eat?"

Greg locked the car without noticing Emily had left her window cracked open. He entered the building and approached the floor broker.

"Well, you did it again." The little man was ecstatic. "You must have a crystal ball. Trad Television is soaring today."

"I want to pick up the check for $3,000 and my stock certificates."

Forty minutes later, Greg had his certificates and collected his $3,000 check. He promptly cashed it in the Citizen's Bank and returned to his car. As he entered the lot, he saw that something was terribly wrong. The grey Ford he'd been avoiding was parked next to his Plymouth. Greg stepped behind a parked car and watched as the Ford pulled away and the driver waved goodbye to Emily. Greg ran forward, unlocked the Plymouth passenger door, and ushered Emily inside. He was too agitated to look in the back seat where a man was lying on his back. Greg slammed the door shut and hurried to the driver's side. He peeled rubber out of the lot and fishtailed onto the wet asphalt of the street.

"What are you doing?" Emily yelled.

"Why was he there?" Greg asked.

"Because I called him! There are too many unanswered questions." Emily moved against the passenger door, her voice a bit hysterical. "I wanted to ask him if he'd recognized you while we were having lunch at the Mission Inn. I still don't know why you're here."

A voice came from the back seat. "Why don't you answer the lady?"

Greg's eyes darted to the rear view mirror. Herman Hoyer was burrowed into the corner. He gripped a small unusual hand gun pointed at the back of Greg's head.

"Who are you?" Greg whispered.

Hoyer shook his head. "Drive north on Main to the river bottom."

"You have a case of mistaken identity. I haven't lived in this town for 30 years. You don't want me."

Hoyer, with the weapon aimed steadily at Greg, ignored him. Greg said, "If it's money you want, take my wallet. There's over three thousand dollars."

Hoyer slowly shook his head.

Greg said, "I can't go with you. I only need two more days in town and then I'll leave, and you'll never see me again."

The river bottom came into view. Hoyer said, "Park by those trees. We'll walk from here."

Greg gazed at Emily. She looked ghastly.

They exited the car. Hoyer had the confident look of a younger man, four inches taller and 70 pounds heavier than Greg. He jabbed Greg in the back to keep him moving forward into the tree- and scrub-covered river bottom.

Greg cast a glance of primal defiance at Hoyer. "Where are you taking us?"

"Not us — you — back to Menlo Park."

Greg spun around and kicked out. The gun flew across the sand. Hoyer reached out with a hand like the talon of a hawk and seized Greg by his neck, jerking him off of his feet. Greg let his body go limp, and then, taking advantage of the momentum, leveraged his body and pitched Hoyer over his shoulder in an awkward Jujitsu throw he'd not used in 30 years. Hoyer, unfazed, executed a professional roll onto his stomach, planted both powerful hands on the ground and sprang upward. His angry eyes bored into Greg's, and Greg swung a roundhouse right fist into Hoyer's temple making him collapse.

Chapter 13

LA HONDA - DECEMBER, 1982

"Well, it certainly looks like you have an event horizon!" Ken pulled away from watching the image through the transparent lead glass window of the modified EM.

Greg had just pushed a tiny piece of brain tissue mounted on a tiny copper EM mounting grid slowly toward the maelstrom with the micromanipulator and the piece of tissue had accelerated as if under its own power toward and then into the maelstrom. It disappeared from the screen. The room roared in its silence.

"You've got a mini black hole," Ken said with a big grin. "You've made scientific history today, old buddy."

Greg said, "This is outside the CIA contract. I don't want these notes in their audited records. Let's hide them."

"Sure, where?"

"In the large storage cabinet in the bathroom."

For several days, they repeated the experiment followed by a variety of measurements on the tissue.

Greg said, "Ken, the brain tissue should have been condensed but not lost weight. It's lost weight."

"I know, it's as if the material just evaporated. One of the most basic laws of science is that matter can't be created or destroyed. So where the hell has it gone?"

"Let's have a cup of coffee and think this through."

They joined David in the cluttered kitchen and selected the two cleanest dirty coffee mugs. Ken poured while Greg expanded on his thoughts.

"It must be all light related. We know that the timelessness of light and its constant speed with respect to any observer lay at the root of Einstein's relativity theory."

David did his two-handed timeout. "Pass that by me again."

Greg said, "The only thing that affects time is gravity, and it does it by lengthening the wavelength frequency of light. Light travels everywhere at 186,282 miles per second. If the frequency of the light at 186,282 miles per second lengthened, then its second also had to lengthen and time would slow down."

Ken interrupted. "In a black hole, the gravitation has become so great that it curves the light trying to leave right back onto itself until the light frequency is reduced to zero. At that point, time stops. There's no longer any 186,282 miles per second because gravity just eliminated the per. Seconds are gone; therefore, time is gone, caput, finished. The singularity is where time ends in a black hole."

David's eyes were sparkling. "So, if we could travel in space to a black hole, time would end for us?"

"Yep," Ken answered, "it would be all over. If you could travel toward the black hole, you would pass through the event horizon without seeing it. But after you'd passed through it, time would start getting slower and slower until it stopped at the singularity."

David said, "If time didn't exist for our space traveler, Dad, wouldn't that traveler be able to travel freely in time?"

"No. After he passed through the event horizon he can't come out."

Greg waved a hand. "Maybe not, In my recent reading, I've come across some interesting material."

"What is it?" Ken asked.

"In the 1960s, it was discovered that there can be two types of black holes: not only a static type that would act as you just described, but also a rotating black hole that carries an electric charge. In a charged, rotating hole, the in-falling observer should be forced away from the singularity along a time tunnel."

David said, "You sound like Rod Serling from the Twilight Zone."

Greg smiled. "An electric charged, rotating black hole may be like the entrance of a tunnel that joins different regions of space-time. Some scientists are referring to them as wormholes. At the end of the tunnel there's an exit they call a whitehole. This could happen because the surface of time and space is curved, allowing an observer to enter in one time and exit in another."

Ken saw the direction their discussion was going and said, "Then if all this theoretical stuff is true, and if a scientist had a static black hole at his disposal like we do, he could convert it into a rotating black hole by adding a charge at the throat of the worm hole. He'd have a time travel machine."

"Bingo!" Greg exclaimed.

Chapter 14

RIVERSIDE - LATE ON THE FIFTH DAY

Hoyer was unconscious and Greg was stunned. He hadn't had a physical altercation in 30 years. He grabbed Emily's hand, and they stumbled back to his car. Greg spun the car about in low gear throwing gravel in a spray.

Emily said, "You told me Hoyer was mistaken about you; but he called you by name, wanted to take you back to Menlo Park. He was so afraid of you he needed a gun!" She turned and faced him.

Greg gripped the wheel so tightly veins stood out on his hands. "After what just happened, I'm going to have to confide in you. I can't do it while I'm driving. Hoyer will be hot on my heels in a matter of minutes." He pulled to a stop at a signal.

"Since he found us at Hogles, he probably knows we're at the Colton Inn. Please Emily, I've got to have two more days."

Emily said, "What normal man would come into town and tell me he was here to change something that happened thirty years ago?"

Greg stomped on the accelerator. "Give me an hour to get relocated, and I promise to tell you."

"I don't know why, but I'll give you the hour." Emily sat quietly staring at him with red-rimmed eyes. "I guess because yesterday was the most beautiful day of my life."

Greg pulled into the Colton Inn parking lot. They walked rapidly to the room and gathered up their belongings. Greg left payment on the bureau and they retraced their steps to the car.

"I've been thinking. Hoyer must have traced me through the car.

He went to Hertz in Riverside and found I'd replaced the Chevy with this Plymouth. I need you to rent a third car in your name at a different agency."

Emily grimaced a nod.

An hour later they were checked into a run-down motel on the outskirts of Redlands, a dozen miles from Riverside. Their rental this time was a De Soto. Greg had left the Plymouth hidden behind a ramshackle garage in an unused alley in Colton. They took turns freshening up in the bathroom's single rust-stained ceramic wash basin. The bathroom smelled faintly of dry, rotting walls.

He next drove them silently through the hills to Beaumont. They entered a neighborhood of splendid, hushed gardens and columned houses. He selected an attractive out-of-the-way restaurant for lunch.

Greg ordered a bottle of Burgundy. The waitress poured two glasses and walked away. Greg began, with an inconsolable expression on his face.

"Emily, the year is 1953. I come from 1983. *I'm a time traveler.*"

Emily upended her glass and the waitress scurried over with a towel and then refilled the glass.

Greg talked nonstop without any interruptions from Emily for an hour. He described his work with Ken and black holes in layman terms and how it accidentally led them to the secret of time travel.

"We were confident that we'd made a tiny amount of tissue disappear because of a change in its relative time. Our next step was to do it with something bigger. One morning Ken brought in a bowling ball and announced this would be the next object we'd send away in time.

"Of course, it was far too big to be put into the electron microscope sample chamber. We needed a bigger vacuum- evacuated area to accommodate larger samples."

Greg spread his arms. "The place to find a large vacuum chamber was at NASA at Ames Moffett Field in Mountain View a few miles away."

Emily asked, "What is NASA?"

"It's the *National Aeronautics and Space Administration* and was founded during the administration of a dynamic young president named Kennedy in the early '60s. He challenged our country to beat the Russians in putting a man on the moon. NASA was integral to his challenge. The race was, on and NASA had the industry build them large vacuum chambers in which to test components for simulated outer space high vacuum conditions.

"By 1983, we'd won the race in space and had put people on the moon. The huge vacuum chambers were obsolete and available for sale cheap. We decided to do it right the first time and purchased a used chamber from Ames that was large enough to accommodate a man. While we were at it, we purchased a couple of used prototype moon astronaut suits designed to protect the wearer in the high vacuum environment of the moon.

"It was a complicated task but we were able to interface the EM on top of the cylindrically-shaped vacuum chamber and focus the electron beam down onto the bowling ball. NASA had provided lead glass observation windows that we could safely peer through. It was a day charged with excitement. We had set the bowling ball adjacent to the focused electron beam. We slowly raised the electron gun voltage and beam current to mini black hole conditions. Suddenly we saw the ball slowly rotate in the direction of the beam. It accelerated and vanished before our eyes!"

Emily recognized that everything about Greg's demeanor was a testimony to his sincerity.

Greg continued. "We were aware of a theory that a static black hole like ours could be converted to a rotating black hole with a wormhole time tunnel. I knew that a scientist named Honjo had accidentally

A Man Beyond Time

accomplished this by adding a positively-charged ion beam at the entrance of his static hole. I rigged up an ion gun on one of the chamber ports and fired ions at right angles into the electron beam.

"We installed another bowling ball late one evening and decided to let the vacuum chamber pump on it all night. Our intention in the morning was to fire up both the electron beam and the ion beam adjacent to the ball and simultaneously raise their voltages to 200,000 volts. We drove up together in the morning to our laboratory. We'd purchased a couple of coffees to go and were sipping them when we got out of my MGA. We were pretty excited about the test we were about to perform. I almost stumbled over it. The bowling ball we'd left in the vacuum chamber of the EM the night before!

"At first we were puzzled, and went into the lab. It hit us then at the same time. The ball we were about to send into time had already arrived back in time and was sitting in our driveway!

"We fired up the test and watched the ball disappear. We retrieved it from the driveway, waited an hour, and reinstalled it in the EM chamber. But, first, we looked out the window. Suddenly, the ball materialized again in the driveway! We had sent it back in time by some 10 minutes. We had successfully invented a time machine!"

Greg ordered two glasses of water.

"That was six months ago by my relative time. In a matter of days, we had learned how to send the ball back almost the amount of time we wanted to by varying the intensity of the ion beam current. We learned to place the ball in time over the short haul to within a couple of minutes in 24 hours. We extrapolated that we could place it within about 10 days over a 10-year time travel.

"We next had to see if a living creature could survive the trip. We decided on a mouse. Ken and I modified the bell jar of a vacuum evaporator we had to perform the reverse role for which it had been designed. It now would hold atmospheric pressure within the vacuum of

77

the EM to keep the mouse alive. As we were loading the mouse into the glass jar we heard a thump in the back yard.

"We ran outside and found the jar. We released the lid and the mouse took off across the yard. We returned to the machine and started pumping down the EM while watching the mouse through the chamber window. Suddenly he started to slide across the bell jar floor putting on his little paw brakes. In the next instant the entire bell jar started to slide horizontally and then disappeared!

"It was time to test it on one of us. I argued that since it was my invention, I should be the one to test it. I also didn't really have any kind of family. We discussed ad nauseam what time period I should go to. Ken wanted me to just go back in time 10 minutes. I ruled that out. I only had the opportunity for one trip, and I would be permanently back in that time. I didn't want to risk what I thought was probably certain death just to be 10 minutes younger! I started thinking of my youth, and it dawned on me that I could do something really good for myself and the person I had killed in 1953."

One of Emily's hands flew to her mouth.

Greg looked down at the table. "I'm here to stop me from killing a boy 30 years ago."

The waitress moved to their table and let them know the restaurant was closing until dinner and asked that the check be paid. Greg paid, tipped generously, and put his arm around Emily. She had become passive and quiet. They walked outside and sat down on an old bench bleached by the hot sun and heavy fall rains. The late afternoon was tenuously resplendent with an illusory, portentous feel to it. Emily's profile was sculptured and softened in the light.

"We discussed at length what impact I might have on the future by going back in time and changing an event. We decided it was a very dangerous thing to do, but I felt it was worth the risk to give the boy back his life. I decided that would be the only event of consequence I would attempt to alter. Ken was also concerned I would return to the same time and place my younger self inhabited. He felt it was one more unnecessary risk piled on top of already insurmountable odds. He ran formula after formula unsuccessfully through his computer trying to deduce what might happen if I actually came into contact with my 30-year younger self.

"After a time, he gave up. I was embarking on the most adventurous and perilous voyage in history. Magellan and Columbus were pikers by comparison to what I was about to do. Our plan was for Ken to adjust the ion gun parameters as best he could to deposit me on the outskirts of Riverside around midnight on Sunday, December 1, 1953.

"We spent a month preparing me for the trip. We looked at the practical needs I would have. For example money and identification. You can't imagine what a problem it was for us to find 30-year-old money. Another problem was to avoid having anything made of materials invented after 1953 — plastic for example, a material you haven't heard of. Both my watch and eye glasses were made of it.

"Well, we finally got me ready. We'd been able to rustle up $2,500 in 30-year-old 100 dollar bills, and being the real rat packer I am, I'd held onto every driver's license I ever got plus my social security card. One of my primary concerns, if I actually lived through the trip, was survival in a world I'd pretty much forgotten. I would have to have a way to generate money. So I went to the library and spent days studying several weeks of 1953 newspapers, memorizing the headlines and most erratic stocks of the day. After that, I felt as prepared as well as I was going to be.

"The night before I was to go, I looked at my wife and daughter engrossed in their exclusive world and felt a tremendous loss of what should have been, but was not. I interrupted them during dinner and mentioned I was going on a trip. It hardly registered. After dinner, I took my old dog Pee Bee for a walk and hugged him goodbye.

"In the morning, I dressed in the used clothing I'd purchased at a Goodwill store that appeared indigenous to 1953 and gathered the few belongings I intended to have with me. I waited for Helen to get up and told her I wouldn't be coming back. She was pouring coffee and didn't say anything for a minute. Then, without turning, she said, 'That's probably the best thing for both of us.'

"I drove to Ken's laboratory. Ken, David, and I had set up cots in anticipation of my trip. I was in absolute terror but didn't let on. That night, Ken and I really tied one on and became extremely maudlin. We were closer than most brothers. He was crying when he said, 'Even if you make it back 30 years alive, I won't know it.' I thought for a moment and answered, 'I'll put a personal in the local newspaper addressed to you for a week. It will say; *Ken, old buddy, I'm alive and well in 1953. Give my best to Pee Bee, Greg.'*

"Ken was to call and ask the paper to photocopy the newspaper for the week of December 1, 1953.

"In the morning, with a ferocious hangover, I put on the NASA space suit, complete with an oxygen bottle backpack, and entered the chamber. After a last hug, Ken closed the vacuum lock and waved at me through the lead glass viewing window. David began snapping picture after picture of me through the window with a wide-angle lens 35mm Nikon camera.

"I was jerked as Ken pushed the start button on the rotary vacuum pumps. Within 30 minutes, the chamber was pumped down to a vacuum comparable to that on the moon, and I was no worse for wear except for being in a highly uncomfortable position. Ken fired up both the electron and ion beams and slowly increased their voltages. I began feeling a gravitational pull in the horizontal — a very strange feeling. It increased until I could no longer maintain my footing and started to fall into the center of the machine.

"In the next instant, I found myself in the path of a speeding car.

The driver had excellent reactions and was able to swerve quick enough to only graze me. For a moment, I was totally confused and then realized where I was: Main Street in Riverside, 1953. Instead of it being midnight Sunday, it was early morning Monday, and I was within a few blocks of downtown, not five miles on the outskirts. All in all, not too bad, eh?"

Emily had grown reflective. "Phil, I love you, but you're absolutely crazy as a loon. I think you actually believe this fantastic tale."

Greg ignored her comment and continued, "Ken and my concern about a meeting with my younger self was portentous. The driver of the car that almost killed me on my arrival was me!"

Emily was barely able to suppress a smile. Greg reached into his pants pocket and pulled out the copies of his recent stock transactions and handed them to her.

"Honey, look at these. Note that I purchased Cinerama Inc. on Monday for $1,800 and sold it on Tuesday for $5,400. I then purchased Gaspe Oil Venture and Nevada Tungsten on Tuesday and sold them Thursday for almost $11,000. I took $7,500 and bought Trad Television. As of this morning it had nearly doubled in price. I also took out $3,000 in profit this morning. In five days I've increased my original investment by 10 fold. Is that the action of a crazy man? On Monday, I'm putting everything into Chemstrand. This stock will soar over the next ten years."

Emily smiled at him affectionately and said, "Just because you're a whiz in the market doesn't mean you're not nuts. I never doubted your intelligence."

Greg said, "Check the stocks. Of all the hundreds of stocks available, those four had the greatest overnight increase. The odds of me being able to pick them are astronomical."

"Greg, I think you've been under too much stress and have had a nervous breakdown. I'm going to call your wife in Menlo Park and tell her where you are for your own good."

Greg emptied his pants pocket change into her hands and said, "Be my guest, but when you find her not listed, call Savannah, Georgia, and ask for the Bettis residence. You may be able to talk with an 18-year-old Helen."

Several minutes later Emily exited the telephone booth with a perplexed expression on her face.

"There were no listings for you or your wife in Menlo Park. I did talk to a young lady in Savannah named Helen Bettis, who thought I was pulling her leg and hung up on me."

"Now are you convinced?"

"Oh, sweetheart, of course not. Your bizarre tale is so threaded with logic, I'm not surprised that in your illness you've subconsciously fabricated supporting evidence."

"Emily, have you ever experienced *precognition*? You probably have, or at least know others who have. It quite often takes the form of a dream in which a disaster is seen taking place — like a plane crash. Later there is just such a crash as the dreamer experienced.

"If not, how about *deja' vu*? A sensation lasting for just a few seconds that what is happening at that moment has happened before. Or *premonition*? A feeling that something important is about to happen. Aren't each of these just a form of people seeing into the past or future? Have you never experienced at least one of them?"

Emily had adopted a tense posture.

Greg continued, "Do you remember when I first took your class I was a real pain in the neck, and you kept me after class? You told me I was intelligent and could do far better than I was but was wasting my time and yours. You called me an irresponsible brat. That was the day I fell in love with you and have remained in love for 30 years. I was ten years younger than you then. Now I'm 20 years older. I used to make it a point to bump into you and have a mini-date at the Four

Cones ice cream parlor."

"How did you know all that? That was just last year," Emily whispered.

"Last year for you, 30 years ago for me. I'll give you some current proof. Today is the sixth day of December, 1953. I've been with you all day. The Evening Press should be hitting the stands within the hour. I'll tell you what the headlines are going to be. There will be three new stories not reported on before tonight's paper. The lead story will be a joint denunciation of Senator Joseph McCarthy and his McCarthyism by President Eisenhower and Secretary of State John Foster Dulles.

"The second headline will be a surprise announcement that Vice President Richard Nixon and former President Herbert Hoover have just returned from a secret visit to Iran. As a result of their meeting, Iran has agreed to reestablish diplomatic relations with Great Britain and resume oil operations.

"The final headline will be the latest war news from Vietnam. The French will claim to have forestalled a new communist offensive in the Indo-China War. Now, let's go get a paper so you can see."

An hour later they watched a newspaper delivery truck drop off a bundle of papers fresh off the press at a news stand on the main street. Greg got out of the De Soto and waited for the newsboy to open the bundle. He bought the top copy and returned to the car.

Emily was ghostly white as she set the newspaper aside. "I don't know how you're doing this."

"Do you want me to tell you tomorrow's headlines? Sundays? Next week's stock results?"

Greg reached into the pocket of his red and white checkered jacket and felt for the De Soto keys. They were snagged by something and he pulled them lose. A penny fell to the floor. He leaned forward and picked it up. As he started to pocket it, he noticed the rear of the penny showed an image of the Lincoln Monument. The rear of 1953 pennies showed an image of wheat ears. He scrutinized it closely and then grinned. "Look what was caught in the bottom of my jacket pocket."

He handed it to Emily. Her hand was trembling when she handed it back. *"Oh my God, you are what you say you are. A time traveler!*

The penny had been minted in 1980.

That night, Emily snuggled tightly into his curled back and asked, "Who do you think Herman Hoyer is?"

"I don't know. But he must be a time traveler. You mentioned a flying saucer sighting yesterday. Didn't you say it was seen at a distance hovering over the Santa Ana riverbed?"

"Yes."

"That's where Hoyer forced me to drive when he said he was taking me back to Menlo Park. I think Hoyer has been sent to find me in a much more sophisticated machine than I designed. I bet the saucer is his time machine. Hell, I bet all UFO sightings, that's what we call them in 1983, have been time travelers!" Greg paused. "I'm still ... in awe ... that I'm actually here with you now ... in 1953."

"Phil, I mean, Greg, I'm absolutely bushed. I can't hold my eyes open any longer."

Chapter 15

THE SIXTH DAY (Saturday)

"Tell me about killing the boy."

They had awakened at daylight and were having an early breakfast in a small chrome plated diner along motel row.

"It started in the high school parking lot after a Bear's game with San Bernardino. Everybody was trying to leave at the same time. A blue '39 Mercury cut me off as I tried to pull into the slow moving flow. I honked, and a Mexican kid stuck his head out of the driver's side of the Merc and told me, 'Screw off!' I yelled back, 'Why don't you make me, grease ball?' And a fight was on. We got out, squared off and traffic came to a standstill. It was over in a minute. I twisted his wrist and threw him to the ground — a throw my best friend, Ralph, and I had practiced out of a jujitsu manual. Not content with having won the fight, I ridiculed his car. We argued about who had the hottest car. It evolved into a drag race bet. Just as we established the place, time, and money, the cops arrived and took us in. The boy, Romo — I learned at the library last Monday that Antonio Romo was his full name — got snotty with the cops. I acted polite, and when my Dad arrived, they let him take me home. Romo's wrist was so badly swollen they took him for medical help and the Press-Enterprise picked it up at the hospital as a small local story. That was Friday night a week ago."

Emily toyed with her scrambled eggs. "Something like that happens after nearly every game."

Greg took a bite and shook his head. "But not with the same consequences. The drag race was scheduled for dusk on Sunday — tomorrow — on an out-of-the-way orange grove service road a mile beyond Highgrove. One hundred dollars was on the line, all the money my gang could scrounge together I was concerned, because even though I only knew his first name, I was pretty sure he was a member of a Pachuko gang from San Bernardino. They were notorious. I took my Strong Street gang of five with me.

"Romo had brought his entire Pachuko gang. The race was to be on the service road lined on both sides with orange trees. We were to go around a bend and cross a single lane bridge with a concrete abutment. The first one across the bridge would win.

"Ralph won a coin toss with the Pachukos and walked out several yards and waved the start. Romo took the early lead but missed his speed shift into third, probably because of his bandaged hand, and I took the lead as we went around the bend. I was half a car length ahead as we came upon the bridge. The tough little guy tried to crowd me over but I wouldn't give an inch as I zoomed over the bridge. He hit the abutment head-on." Greg wiped his eyes with the back of his hand.

Emily had stopped eating. "Was there anything you could do?"

Greg shook his head and looked down at his nearly full plate and pushed it away. "I should have backed off and given him the road. His old Mercury exploded. I saw him banging at the window with his bandaged hand with flames all around him. The morning Enterprise carried the story of an anonymous call reporting the accident. No witnesses came forward because I was the only one who knew what happened."

Emily asked, "What did you do next?"

"I enlisted in the Air Force as soon as the recruitment office opened the next morning and asked to be processed immediately. I was put on a bus that evening to report to basic training the following day in northern California."

"And that was all?" Emily asked.

"I came home when mom was sick and again to bury her."

Emily pushed her plate away. "How do you propose to change the outcome in Highgrove?"

"The drag race is tomorrow at dusk. I know Greg is off with Ralph this afternoon seeing a new cinemascope movie called The Robe. Ralph's driving. Greg won't touch his car again until it's time to leave for the race tomorrow evening. I intend to disable his car this afternoon while he's gone."

Emily asked, "What's next on our agenda?"

We'll visit mom and dad this afternoon."

"How do you feel about that?"

Scared. The last time I saw both of them was to bury them. I'm afraid I'll break down in their presence. You may have to do most of the talking."

Emily said, "We better buy you a pair of sunglasses to wear to disguise yourself a little."

Twenty minutes later, Greg exited a store wearing a large pair of sun glasses. "How do I look?"

"Silly, but they do the job."

Greg became increasingly somber on the drive to Riverside. Emily reached across the seat and took his right hand in her left. Greg said, "How are you dealing with all this?"

Emily replied, "I'm overwhelmed by it."

Greg pulled into Nicolino's, an out of the way Italian restaurant several miles east of Riverside.

"We've got a couple of hours to kill until we visit my folks. Let's have ourselves a bracer."

With drinks in front of them, Emily tried to lighten the gloom that had settled over Greg.

"Greg, exactly what is time?"

"Just one damn thing after another. Do you know why God invented it?"

"No. Why?"

"So everything wouldn't happen all at once. Be too damned confusing — just a minute."

"What is it?"

Greg leaned forward and whispered, "My old buddy, Ralph, just walked in with my high school sweetheart. They've slipped into the booth right behind us."

"Are you being cuckolded, dear?" Emily asked sweetly.

"Shush, let me listen." Greg leaned back and pressed an ear against the seat. After a time he leaned forward again and whispered, "The SOB is trying to convince her to dump me in favor of him, and she's actually waffling. Says she has doubts about my long term commitment! Ralph is giving her assurances about his."

"You probably wouldn't have been any better off with her than you were with Helen." Emily's eyes twinkled.

"Let's get out of here." Greg stood and helped Emily out of the booth. He looked down at an unsuspecting Ralph's peroxide bleached hair swept back into a duck's ass. Ralph looked up. His nose was bandaged. Greg glared at him.

Emily grabbed Greg's hand and started swinging it as they exited.

Greg turned onto Strong Street, down shifted into second and slowly cruised the street. Old two bedroom wooden houses, mostly of a 1930 vintage, lined both sides. No sidewalks. Dogs were sleeping or playing in most of the yards.

He pulled into his parents' driveway and parked beside the gray 1940 Ford. He reached under the driver's seat and stuffed something into his jacket pocket. They walked to the front door and knocked. His mother Kathryn opened the door. Greg thought he was going to faint.

Emily introduced them as members of the juvenile division of the local police. Kathryn opened the screen door and invited them in. They entered the dimly lit interior. Greg's father was reading the latest issue of Colliers magazine under a small Chinese lamp. He removed his glasses with the thumb and forefinger of his right hand. The remaining fingers were splinted and bound together. He looked at Greg and Emily. A long forgotten odor of home assailed Greg — a musky mixture of stale pipe tobacco and coffee. Greg's throat was so dry he couldn't swallow.

"This is my husband, Edward," Kathryn completed the introductions. "Please have a seat. I was about to get us some coffee. Would you like some?"

Emily said yes. Greg managed to nod.

Kathryn moved toward the kitchen with a lurch, swinging one leg noticeably shorter than the other in an arc as she walked. She returned with the coffee and asked, "What can we do for you?" She looked at Greg. He ached to hug her. He answered from a mouth filled with wads of old cotton. "Your son was in a fight a week ago with a Mexican boy at the high school. We have reason to believe they're going to meet again tomorrow night and escalate it into a gang fight."

"I don't know what to do about that boy," Edward interrupted. "He's been in and out of trouble ever since he entered his teens. Lord knows I can't control him."

"We'd like you to make him stay home tomorrow."

"Can't do it," his father said. "The boy is too head strong for us. Can't you find out where this fight is going to be and have your officers break it up before it starts?"

Greg sipped his coffee and observed his father. Edward's demeanor was fraught with lassitude. Greg had his second ploy prepared.

"The fight's not the only reason we're here. We believe your son has been burglarizing Reids Automotive Wrecking Yard at 10th and

Lime."

"Nothing surprises me anymore."

"Stop it, Edward." Kathryn interrupted. "He's not a bad boy. Please don't give these people the wrong idea."

She turned to Greg and continued, "Greg is a fine young man. A bit wild, but not a mean bone in his body. It's not his fault we had him so late in life. I can't believe he's involved in a burglary. Do you have any evidence?"

Greg gazed at her with profound love and lied. "All used engine parts at Reids have had an R scratched on them. Boys usually burglarize for very specific engine needs of their own. If I could raise the hood on your son's car I could tell in a matter of minutes if he has stolen anything."

"Sure, go ahead," his father said.

"Edward, are you sure?"

Greg looked at his mother in admiration.

"Kathryn, if he's started stealing, we may as well find it out now. No sir, you go right ahead and look under his hood."

Emily said she would stay and keep the Philips company while he was looking. Once outside, Greg peered through the open living room window and saw that Emily had his parents engaged in conversation. He raised the Ford's hood and quickly removed the distributor cap. He switched it for one in his pocket. They were both old and greasy and looked identical. After a few minutes he reentered the house.

"We were wrong about your son, ma'am. There were no stolen parts on his engine. I'm sorry we bothered you."

"Thank heavens!" Kathryn had a sad exhausted air about her.

"You seem to care for your son very much, Mrs. Philips," Greg said.

"Yes I do. I worry sometimes whether he knows it or not. We were much too old to have him, and I'm afraid he's had to grow up on his own. We're such an old fashioned pair, but yes, I love him very much. Just not much good at showing it, I guess."

"How about you, Mr. Philips?"

"What about me?"

"Do you care for your son?" Greg asked.

"I'd say that question is a bit out of line, wouldn't you?" Edward squirmed around in his chair and glared.

He settled back into a somnolent state. Kathryn took his good hand in hers.

"What did you do to his car?" Emily asked as they drove away.

"I switched distributor caps. Last Tuesday, I bought a used distributor cap for Greg's Ford and spent Thursday morning disassembling it and removing a critical wire. Externally it looks okay, but there will be no electricity passing through it again."

"Your mother is really quite lovely. You hadn't mentioned she was crippled."

Greg's eyes brimmed in tears. "She wasn't crippled. I don't know what's going on."

They gazed at each other uneasily.

"Oh, Greg, I'm so sorry. This is all so unreal! What was that all about with your questions at the end?"

"I always thought both my parents didn't give a damn about me. I was profoundly affected when my mother defended me. I had to ask her if she loved me."

"She obviously does," Emily said as she squeezed his arm. Greg said, "Let's get a hamburger. I think I'd feel better with something in my stomach."

Greg parked on Seventh, and kept his eye out for Hoyer. "Look at that sign, Emily." A colorful banner over the little McDonalds hole-in-the-wall hamburger stand proudly announced,

Over 2 million hamburgers sold to date. Placed side by side, they

would reach from here to San Francisco.

"This is one of just two McDonalds in 1953. The other one is in Downey. They're both little joints. In 1983 they have become the biggest hamburger chain in the world. They've probably sold enough hamburgers now to reach to the Milky Way."

Emily shook her head in disbelief.

Seating in this early McDonalds was at the counter only. Eight stools. They were served by a single fry cook. The hamburgers were 15 cents each. They weren't lucky and didn't receive the promotional red mark on their receipt. Greg couldn't get his mind off the fact that his mother was crippled. They ate their hamburgers driving back to their motel.

Greg unlocked the door and reached for the light switch.

"Hello, tough guy."

Hoyer was seated at the metal desk with his strange weapon pointed at Greg.

"How did you find me?"

"I just waited at your parents'."

"You're a time traveler aren't you?"

"I'm here from the year 2030."

"You're able to travel forward and backward in time?"

"Yes, thanks to you."

"Thanks to me?"

"You are venerated where I come from as the father of time travel."

"Are you serious? Why have you chased me down?"

Hoyer waved the gun at Greg to sit down on the far side of the bed with Emily between them.

"I'll take the time to tell you so you'll understand how you are endangering mankind's future even as we sit here."

Hoyer rested his elbows on the desk. The gun stayed trained on Greg. "I'm tracking administrator for the National Time and Space

Agency. In other words, I'm the chief time travel cop. Shortly after we worked out the fine issues for forward and backward time travel, everybody wanted to try it. We put a moratorium on all travel until our scientists could study what impact time travel might make on the future. Their findings were disturbing. We found every time traveler who goes into the past causes a split in time. One branch carries on as before, but the other branch with the traveler in it will contain differences caused by his presence. The more robust of the two will determine the future! A careless traveler can cause irreparable damage to all of our futures. Congress voted in 2015 on only allowing time travel astronauts with extensive training to conduct the most limited of time travel. Our first charter was to clean up the early travel messes, including yours. You're the last one, and this is my last trip out."

"What if I don't want to go?"

Hoyer said, "You have no choice. I have authority to use as much force as I need. Rest assured I'll be happy to use it."

"How did you find me?"

"We looked at the settings on your primitive machine and could make a calculated guess, give or take 20 years. We've also found through surveys that the majority of people will always choose the time and place of their youth. They all want to see their deceased parents again. You were no exception."

"Am I the father of time travel? I think it's possible that both a Canadian and a Japanese produced an electron microscope-induced black hole before me."

"Pratt and Honjo. They didn't know what they'd accidentally produced and traveled unintentionally in time. It nearly killed both of them.

"You, on the other hand, knew precisely what you were doing and kept copious records that are now in the Smithsonian Institute."

Greg's eyebrows raised in astonishment. "What happened to Pratt and Honjo?"

"We sent a fleet of Time Astronauts into every century. It took several years to locate them."

"A fleet of saucers?"

Hoyer raised his eyebrows.

Greg said, "I saw strange tracks in the river bottom sand the day you took me there. Where'd you hide your saucer?"

"At a pre-selected remote longitude and latitude. I can call it back whenever I need it."

"When I traveled in time here a week ago," Greg commented, "my electron microscope didn't accompany me."

"We made many improvements to your primitive design, including having it travel in time with us." Hoyer said.

Greg sat down on the edge of the bed. "Let's get back to my predecessors. What happened to the Canadian, Pratt?"

"He was my first lost time fugitive. When my astronauts finally found him, they called me in. Pratt didn't want to return to 1947. I broke his neck trying to strong arm him into the saucer, but I was able to deposit his body back in his own time."

Greg frowned, "How about Honjo?"

"Another hard core. He'd landed back in the days of the Shoguns in the 17th century. With his knowledge of modern science, he ingratiated himself with the current regime. By then, I was more sophisticated in my persuasion techniques. I warned Honjo I would travel back to his youth and expose him to a debilitating disease."

"And?" Greg asked.

"He didn't believe me."

"What happened?"

"Rickets."

"What happened to my mother?" Greg's voice lowered an octave. Hoyer said, "I decided to give you a little advance persuasion. I dipped into the time when she was two and exposed her to polio."

"You bastard!" Greg leaped to his feet.

"Careful tough guy, or I'll shoot the young lady."

Greg's face was perspiring, "Why did you hurt my mother? I'm the one you want."

"I figured it would save time to show you in advance that I'm invincible."

"Weren't you tampering with the future when you subjected my mother to polio?"

Hoyer waved the gun for Greg to sit on the bed again. "I'm required to do a documented computer analysis of any potential time splits my visits might cause. I can cause no time split that will have any effect on the future. In your mother's case, she still married your father at 38 and had you at 44. Like in Honjo's case, no change in the future except for her own discomfort."

Greg sat down. "Well, I'm not going anywhere. Go back to 2030, and tell them I'm not coming."

"You are my final fugitive. Once you're back in 1983, the primitive world will have seen the last of my UFO fleet, and I'll be promoted to Director of the Time Travel Agency, reporting directly to the president. I can't let you stay here and jeopardize that."

"There's not much you can do about it." Greg said.

"Do I have to add more miseries to your mother's life?"

"Leave my mother alone." Greg paused. "I'll come with you if Emily comes."

Hoyer shook his head. "It would be just as risky to have her left in future time as to leave you in the past."

Greg looked at Emily with an overwhelming feeling of despair. He said to Hoyer, "I chose this period in time because of something I must do within the next 24 hours. I won't consider leaving until that's done."

Hoyer asked, "Do I have to return again to your mother's youth."

Greg ignored him. "We both know you can't drag me back and you'll have a problem if you kill me. I'm staying for another 24 hours. If you hurt my mother, I'll do something devastating to disrupt this time."

Hoyer said, "If I give you the extra time will you promise to come with me without a hassle in 24 hours?"

Greg turned to Emily. "Do you have any ideas I've missed?"

Emily's eyes were moist. She shook her head. Greg said, "You've got my promise, Hoyer."

"I'll see you here tomorrow night." Hoyer walked out.

Greg and Emily didn't sleep that night. They made plan upon plan that they knew was fruitless. By early morning, the subdued light of winter suffused the room. Greg got up and signed over his stock certificates to Emily.

Chapter 16

THE SEVENTH DAY (Sunday)

In the late afternoon, they dressed and drove to Highgrove. They parked the De Soto on the main road in the tiny hamlet and walked into the orange grove. Greg found the Sunkist truck loading area and took them to a vantage point he had selected during his visit to the site the previous Thursday. They waited for dusk.

A caravan of three old cars of different makes, shapes and sizes arrived first in a cloud of dust. A dozen young Mexican men got out and milled about. Greg recognized the '39 Mercury and saw Antonio Romo. His thick black pompadour hung loose in a romantically Latin wave across his eyes. Greg waited anxiously for his Strong Street bunch praying that young Greg had not been able to diagnose what was wrong with his car and repair it in time.

He heard a Ford's eight cylinders blasting exhaust through a system of dual mufflers. He strained to see if it was his old Ford as the car skidded to a stop. It was a '34 Ford sedan. Four doors opened at once. Greg sighed. It was Ralph's car. A second concern hit him. Was young Greg driving? He felt immense relief as Ralph came out from behind the wheel. Young Greg came from the passenger side. All attendees became very quiet. Old Greg was mesmerized looking at himself and this Neanderthal period of his youth.

Young Greg approached Antonio and said, "My car's dead. I think it was sabotaged. I can't start it. I tried for an hour." Antonio postured and held out his good hand palm up. "No race? Then you owe me."

Young Greg held up a forefinger. "But, we're not backing out of the bet. My buddy, Ralph, has agreed to take it in my place if you're up to it. It's better for you. His car isn't as hot as mine."

Antonio huddled with the Pachuko gang and decided to accept.

"We're ready when you are." Young Greg said, "I'll flag the start, okay?"

"No problema," Antonio answered.

"Then let's do it!" Young Greg exclaimed with perspiration glistening on his forehead.

Young Greg walked ahead 50 feet with a flashlight. Ralph and Antonio had parked parallel on the asphalt service road with their engines running. Respective gang members crowded around on either side of the road. There was palpable tension in the air. Final rays of sunlight blended with the light from a full moon. Together they illuminated the scene as they filtered through the green and gold of the orange trees.

"Are you ready?" Young Greg yelled above the engines.

Both cars honked in unison. Young Greg switched on the flashlight and held it in two hands aloft dramatically. He dropped the flashlight between his knees and the two cars spun their wheels. A stench of rubber drifted across to Greg and Emily in their hidden position on the upper dock. A second squeal of punished rear tires roared in the night as the two drivers slammed into second. Romo pulled a little ahead as they rounded the bend and out of sight. A few moments later, they heard the sound of a horrendous crash. A orange glow of an erupting inferno appeared over the trees. Greg squeezed Emily's hand painfully. Both gangs suddenly went totally quiet. After a couple of more minutes, the 34 Ford returned around the bend.

Greg looked at Emily in dismay. His plan to save Antonio's life had not materialized. The split in time change of events because of his interference was minimized. The result of the race was the same. Only Ralph now had the curse, not Greg.

They drove in heartbreaking silence back to the motel. Hoyer was standing outside their door. Greg held Emily for the last time, and kissed her tenderly. He then emptied his wallet and pockets and gave her the $3,000-plus dollars he still had and the car keys. He turned and walked away with Hoyer without looking back in the calamitous throes of his loss. He hadn't had such a feeling of remorse since he'd caught the Greyhound for Air Force basic training 30 years earlier, and this loss was an order of magnitude worse.

Chapter 17

MENLO PARK - 1983

G reg looked around. The night was awash in moonlight. He was in the front yard of his house in Menlo Park. He could see the muted flickering colors of a television through gossamer curtains covering the front window. It was early evening in 1983. He stood despondently looking at the front door.

He remembered the ride to the river bottom with a triumphant Herman Hoyer. A device that looked like a garage door opener called the saucer back from storage, and they entered. Hoyer put Greg into a seat beside him and gazed at a color monitor displaying control parameters. He punched in selected data. Greg felt the same unusual feeling of falling into horizontal gravity. The entire saucer seemed to implode around him toward the center. He'd blacked out until finding himself in his yard.

Greg noticed Ken had returned his MGA and left it in the driveway. He shuffled forward and rang the bell.

Helen opened the door. "So, you're back."

He walked past her toward his room.

"Just a minute. You can't disappear for a week and not tell me where you've been."

"On a business trip."

"Wait a minute, there's something you need to know. Carrie left the door open and Pee Bee wandered out into the street and got run over. We had him put to sleep."

"Is Carrie here?"

"Off to Tahoe with a new boyfriend."

Greg closed the door and fell to the bed in a cloudy gloom of despair and slept. In the morning, Helen knocked on his door and heard Greg say, "Leave me alone."

An hour later, she called through the door. "Sleeping Beauty, you've got a call from a guy named Southwell."

"Hello, Jeff."

"Greg, what's going on with Ken?"

"What do you mean?"

"I've been trying to call him at the lab and home for days. Do you know where he is?"

"No, I've been out of town."

Jeff said, "I'm a little worried about him. I was going to take the morning off and drive over. You're so much closer, can you check on him?"

"Of course. I'll run up to his lab now and call you back."

Greg dialed Ken's number and listened to it ring. After a quick cup of coffee, he drove into the hills. He pulled into Ken's secluded laboratory. At the front door, a pungent odor greeted him. He knocked to a silent house, tried the knob and found it open. The odor of putrefaction overwhelmed him; a powerful, sour smell, which dizzied him in the first breath. Something was horribly wrong. He moved quickly across the entryway and opened the door into the hallway leading into the kitchen. He froze. Dave lay heaped on the floor on his stomach. His head twisted abnormally around with very dead eyes staring at Greg.

Greg took out a handkerchief and covered his nose quelling an emetic queasiness. He stepped over Dave and moved into the kitchen. The horror of the scene paralyzed him. Ken's head was flopped onto his chest. He was tied in one of the kitchen chairs with shipping straps left from the delivery of the EM. Greg reached out and slightly raised

Ken's head and looked into eyes reflecting such total misery and despair it caused him to gasp. Cigarette burns covered his face. The degree of rigor mortis indicated he had been dead for days.

Papers were strewn about the floor. Cabinets were tipped with drawers left open. A robbery that got out of hand? But why had Ken been tortured? Greg looked into the EM room. The EM vacuum lock was wide open. Files were pulled open and their contents thrown about.

Greg's mind was racing with question. Was the assailant looking for valuables, money? Why wouldn't Ken simply have given them what they wanted? Why was torture necessary? Greg staggered out the front door, took several deep breaths and returned. He was immediately racked by a wrenching bout of tears. He returned to David and gently rolled him over. Dave's left arm flopped out from under his body. Its hand was clutching something. Greg forced the hand open and unfolded an envelope. Inside was a photocopy of a single newspaper clipping: *"Ken old buddy, I'm alive and well in 1953. Give my best to Pee Bee, Greg."*

The wadded envelope was postmarked ten days ago. The return address was the Riverside Press-Enterprise. Greg stared at the photocopy for several minutes. He thought, "Hoyer lied when he said he'd looked at the settings on my machine and was able to calculate the time I'd traveled to. He hadn't been able to find me any better than he had Pratt and Honjo. He was anxious for quick success and tortured Ken to learn what time I'd gone into."

Greg laid the newspaper clipping on Ken's desk and started putting the scenario in place. "David pulled in that morning, picked up the mail and was reading my newspaper clipping as he entered the kitchen and found Hoyer with his dad. Hoyer held David at bay while he read the clipping. David attacked him and in the struggle grasped the clipping. They fought into the hallway and Hoyer broke David's neck. He then returned to the kitchen and broke Ken's. He set the scene to look like a drug massacre but forgot one thing in the heat of the moment. He forgot to retrieve the clipping.

Greg sat with Ken for an hour holding his hand before wiping his eyes and murmuring, *"Ken, I'm going to kill him for you."*

Greg's EM was waiting to be fired up, and there was still a second astronaut suit. He was going traveling in time again. He thought, "I can return to Emily! When? 1953? Hoyer would come immediately. It took him years to locate Pratt and Honjo. It'll take him longer to find me if I go randomly into time , No, it has to be a time with Emily. I'll need time — within that time — to prepare for Hoyer's arrival. He has no advantage over me in this game except his ability to travel both backward and forward in time; I'm limited to travel backward. Or am I?

"I'm going to make him play by my rules and then kill him. My God, he can arrive any minute. What has to be done? How can I operate the EM while I'm inside it?" He answered his own question. "The rheostats and switches I need are here. I can wire them in."

He placed a call to Bob Buchanan.

"Greg, how the hell are you? Have you found your holograph yet?" Buchanan's voice boomed over the phone.

"No, but I have a question that you know the answer to."

"Try me."

"When did Hitachi 200 KV EMs first become commercially available?"

"The Model HU-200A first sold in this country in 1968."

Greg next placed a long distance call to the Riverside Press-Enterprise newspaper. He asked for someone with access to previous year's news stories. He found himself being passed from department to department. The fourth party who came on the line was Ann Hernandez.

"Miss Hernandez, Can you help me?"

103

"I've been here for 20 years, maybe I can."

"My name is Dave Hoard. I'm a freelance science writer in Northern California. I'm doing an article on the frequency of UFO sightings in different localities around the state."

"Sounds interesting. How can I help you?"

"Riverside has an above average number of sightings over the years. Could you spot check for sightings on a monthly basis over the years 1953 through 1973? I'd be happy to pay the newspaper or you directly for your time."

"It sounds like fun. Those years are all on microfilm. Sure, I'll do it for nothing. Twenty years in circulation can be drudgery. When do you need it?"

"Could you do it this evening?"

"I guess so. Promise me an autographed copy of your article?"

"Of course. Could you call me collect and tell me from your notes on what dates there were sightings?"

"Sure, what's your number?"

Greg gave her Ken's number and warmed up a soldering iron. By 10:30 p.m., Greg had wired in the rheostats and vacuum pump switches he needed to have inside control of the machine.

At 11 p.m., the telephone rang with a person-to-person collect call.

"I've got what you want, Mr. Hoard."

"Terrific, Miss Hernandez. What does it look like?"

"There's an interesting trend that might support your story. There were sightings almost on a monthly basis from 1953 right through 1969, and then suddenly they stopped altogether.
Not a single sighting between 1969 and 1973. Is this good information for you?"

"It's exactly what I'd hoped. Please list the dates." He reached for a pencil and pad. "Uh huh, uh huh, thank you so very much, Miss Hernandez."

He hung up the phone. "It looks like I'm as ready as I can be," he thought. He quickly wrote a note to Southwell:

Jeff:

In the process of looking for holographs within the EM on your CIA contract, Ken and I discovered a mechanism for time travel incorporating the EM. Ten days ago, with Ken's assistance, I traveled back in time 30 years, where I spent the next seven days. A time travel control policeman from the future year 2030, named Herman Hoyer, found me within three days and coerced me to return to 1983 four days later.

I visited Ken this morning and found him and his son tortured and murdered. I am sure the murderer was Hoyer. Hoyer lied about his ability to deduce the year I'd traveled to. He learned it only after torturing Ken. I know this will sound bizarre, but look at the attached newspaper clipping from the Riverside newspaper dated 1953. I placed that personal for Ken ten days ago. The wording was by prior agreement between us. This arrived in the mail the day Hoyer was torturing Ken. Hoyer used the information to find me immediately in 1953.

I'm going back into time. Hoyer will shortly be aware and realize I know he's a murderer. He'll follow to dispose of me and cover his tracks. My only chance is to escape into another time in ancient history he won't suspect. By his own admission, it takes him years to just cover a century. Please see that Ken and David get proper burials.

Greg Philips

Greg made copies of the note and newspaper clipping on the Xerox machine and left the original in the EM room. "It'll be Hoyer who finds the original of Southwell's note, not Southwell," Greg thought. "Hoyer won't know if my note is ingenuous or not. He knows I'm aware he'll expect me to try to rejoin Emily. And he spot checked Riverside through 16 years on a monthly basis according to Miss Hernandez. But he didn't find me or catch me until 1969. I'll have a year in Riverside before he arrives. That's all the time I'll need to prepare to kill him."

He picked up Dave's 35 mm Nikon camera and removed the roll of film. He took it and the tightly folded copies of the note and clipping and buried them in a large bottle of aspirin from the bathroom storage cabinet. He next scanned through Ken's
telephone chart of numbers and called Jeff Southwell at home.

"Sorry to wake you, Jeff, but I'm about to leave on a trip."

"That's alright. What's up?"

"Ken is tied up at the moment, but would like you to drop by the laboratory in the morning and pick up some film he thinks you'll find interesting. It's hidden in a bottle of aspirin in the bathroom storage cabinet."

"I could come over now."

"I'm leaving now. The morning will do fine. Goodnight, Jeff."

A few minutes later, he was dressed in the astronaut suit and inside the EM chamber. Greg activated the remote controls he had soldered into place and adjusted the EM parameters. When they reached the appropriate values, he felt the now familiar tug of horizontal gravity pulling him off his feet.

Chapter 18

NTSA - 2030

T he tall young man was dressed in a powder blue uniform tailored in a style reminiscent of the rakish Nazi SS uniform from the first half of the last century. He paused at near attention in front of an imposing seven-foot oak door and polished the toe of each of his boots on the back of his trouser legs. He was about to enter the office of the director of the National Time and Space Agency. NASA had occupied the facility from its founding by President Eisenhower in 1958 until the first successful round trip time travel in 2014. The 2014 president, Anne Gallagher, had renamed the agency NTSA and given it a new charter.

Before stepping in front of the scanner, the handsome young man wiped a frown from his face. The door swung open and Cody Webster walked briskly up to the director's receptionist, an attractive brunette in her early 30s.

"Good afternoon, Webster. Go right in. The director's expecting you."

A second door swung open automatically and Webster entered the inner office of Director Hoyer.

"The son-of-a-bitch is gone again." Hoyer said.

"Philips?"

"Of course. Apparently within days after I deposited him back in 1983. He used his old prototype microscope again. We should have destroyed it," Hoyer said

"That would have been like destroying Lindberg's plane or Noah's arc," Webster responded.

Hoyer pulled a package of cigarettes from a desk drawer and lit up. "I'm sending you after him. This time you'll disable his microscope."

"Do you know what time period he went to?"

Hoyer inhaled deeply. "That's for you to find out. All we know for sure is that his prototype can only send him back in time." He blew smoke out both nostrils. "So that cuts your search in half."

He laughed and then pointed his cigarette at Webster. "Get him. Get him even if you have to bring back a dead body. But get him. I'll be responsible."

Chapter 19

RIVERSIDE - JANUARY, 1968

Greg found himself on his back in a field. He rolled over, disoriented, looked around, and quickly removed his astronaut suit and carried it toward a two-lane asphalt road. Large buildings were visible in the distance. It was the Kaiser Steel mill several miles outside of Riverside. He climbed a fence and threw the astronaut suit into a slag pile. Greg had no money or identification, so he started walking toward Riverside. The sun told him it was early morning. A pickup truck heading in the right direction sped toward him. He put out a thumb. A redneck driver appraised him and roared past.

He walked along deep in thoughts of Emily. She'd be 43 now. Married? Sick? Alive? He picked up the pace. "I'll go directly to her duplex."

It was early afternoon when he reached her street and walked past the manicured shrubs up the drive to the large front porch. His upper lip was moist with nervous sweat as he rang the bell. During the long walk he had paid the ultimate homage to the fear of not finding her. No answer. Don't panic. It's probably a weekday and she's at school. Does she still live here? There was mail in her box. He looked at it, all addressed to Emily Johnson!

"Thank Heaven, she's still in Riverside and unmarried."

On, Main Street he stopped a middle-aged lady pulling a shopping cart, and asked the time.

"2:15" She said.

"And could you tell me the date and day of the week?"

She looked at him suspiciously. "Friday, the 13th of January, of course!" She scurried away from him.

At the arroyo he paused. "I wonder if she still walks home this way? Shall I wait an hour and see? No. I can't wait." He walked on into the school. He was stunned to see the name plate.
Riverside Junior College. No longer Poly High School. He stopped a boy in his late teens and asked, "What happened to the High School that was here?"

"It moved out to Central and Victoria years ago."

"Did you attend it?"

"Yeah, why?"

"Was there a Physics Teacher named Emily Johnson?"

"No, but a Miss Johnson has been teaching here in the Junior College for awhile."

"You mean she's here?" Greg's heart was racing.

"Right down there in Room 101." The boy nodded his head.

Greg floated toward the room and looked through the door window.

Emily's diminutive figure was pacing in front of the class as she lectured. Time had encircled her softly, enriched and deepened her beauty. She was dressed in impeccable white with a cowl neck knit top and silk kerchief draped over one shoulder. A streak of gray swept through her honey-colored hair. Her nails were richly translucent and sculptured into small white half moons. For Greg, it was pure empyrean magic. She was still the most lovely woman he had ever seen.

She turned and gasped in mid-sentence as she saw his image through the frosted glass door panels. She stood transfixed, unable to move. After several moments, she said something to the class and the students started gathering their books and moved through the door.

Greg stepped aside and overheard several of them muttering, "What was that all about? She looked like she'd seen a ghost."

As the last student exited, Greg walked into the room. Emily fell into his arms and clung to him like a limpet. They kissed long and hard. Emily leaned her head back and looked up into his face.

"You haven't aged at all."

Greg said, "You're more lovely at 43 than you were at 28."

"Why have you taken so long to come for me? I've walked home by the arroyo every school day looking for you for 15 years."

"Why did you think I'd come?" He leaned back holding her shoulders. "When I left you, it was to permanently return to 1983. Herman Hoyer made that clear."

Emily replied, "Because a month after you were gone a young man named Cody Webster stopped me by the arroyo and asked if I'd seen you. He said he worked for Herman Hoyer, who was now the director of the National Time and Space Agency in 2030." She snuggled again into his arms. "Webster interviewed your wife in 1983. She told him you'd disappeared for a week and then returned depressed and suicidal. You then disappeared for good. She was using desertion as grounds for divorce. Hoyer didn't believe you'd committed suicide and sent Webster back in time looking for you."

Greg said, "So Herman got his promotion and has others now doing the dirty work. Has Webster been back since?"

"I think I've seen him at a distance several times over the years, but he hasn't approached me again. Why did you wait so long to come for me? I've missed you terribly. I've waited for you long after my common sense told me I should have given up."

"I couldn't come any earlier. This was the first period in time I felt it would be secure for us to be together. Let's go home."

They walked hand in hand, talking of their divergent lives. They felt free, protected with restored tranquility. The weather had turned chilly. Clouds were filling the sky with subtle shades of pink, mauve, and magenta, changing imperturbably with each passing moment.

It started to sprinkle as they entered her duplex. Once behind the front door, they hungrily undressed each other. Emily was more buxom, softer, and more at ease with her nakedness. Afterward, she prepared a plate of sliced apples, oranges and avocados. She lay in the crook of his arm, as he sat up in bed feeding her pieces of fruit.

Greg asked, "Has there been anyone else?"

"There never was anyone else for me. If you hadn't come today, I would have gone on waiting for you until the day I died."

"Helen was right. I was so depressed I could have committed suicide. There has never been anyone but you for me."

"Why did you wait so many years? How can you escape Hoyer even now? I thought he was invincible when you left."

"So did I. He made us believe he could find me in any time whenever he needed to. He was lying."

"Won't he be coming after you?"

"Yes, or maybe his henchman, Webster. One or the other is going to arrive in about a year. We're safe until then as long as I keep a low profile."

Emily snuggled closer and asked Greg what had happened to him on his return to 1983, and he told her.

Emily sat up in the bed. "You must not have been back in your time a week before you traveled again. That's why you haven't aged! While I've waited 15 years for you, you've only waited a week for me! I'm 15 years older, and you're still the same age!" Emily's comprehension kicked in.

"I was too old for you last time. Now, I'm only five years older. Just about right." Greg smiled.

"Why did you wait?"

"Honey, if I had come back anytime in the 1950s Hoyer would have found me in a matter of days. I had to select a time that would throw him off."

"Why did you select this year?" Emily said.

"This is the first year that 200 KV EMs are being offered for sale commercially." Greg shifted in the bed to a sitting position with his hands behind his head.

"I got a good look at the control console of Hoyer's saucer as we left 1953. There was an EM in the center of it to form the mini-black hole. The entire saucer imploded in on itself as we entered the hole, but not before I got a pretty good look at things. I think I can copy for two-way time travel. I just need another 200,000 volt EM."

Night had fallen, and the faintest wash of moonlight leaked through a curtain. Emily snuggled to his chest, listening.

"Starting in the morning, I've got my work cut out for me. I need to make a lot of money fast. Buy an EM. Rent a laboratory. Do you have any money I can borrow?"

Emily raised herself, her face brightened — the quintessentially sensible school teacher. "I have a lot of money. The day after you left I remembered your plan to invest everything the following week into Chemstrand, and that's what I did. The $3,000 and the Trad Television stock. It's grown to over a million dollars."

Greg squeezed her to his chest. "May I borrow some of it?"

"It's your money. I just wanted to do what you'd want me to. The only luxury I spent anything on was a car."

"You're wonderful. Let's get to sleep, I'm going to start putting a plan together in the morning."

"What is your plan?"

"Let me sleep on it."

The morning ray of light through the Venetian blinds spread across his face, and he leaped out of bed with a feeling of serendipity. Greg was with Emily and was invincible. She had the money to finance his incipient plan. His sorrow for Ken had been replaced with a deep resolve for revenge.

He yelled from the shower, "Do you have a spare tooth brush?"

"No. I don't have men drop in to spend the night," Emily yelled back while scrambling a half dozen eggs with cheese.

"Now tell me your plan," she said as she poured the coffee.

"It's in two parts. The first is how to permanently hide us from the time travel agency. From what Hoyer described, the government of 2030 is paranoid about keeping their people in their own time for fear of possible time splits. By now, Hoyer and his crew know I've traveled again. They'll concentrate looking for me in time before 1983."

Emily asked, "Why?"

"Because they know my machine only worked sending me back in time. We know Webster spot checked here for me from 1953 through 1969. He either found us in '69 or started looking elsewhere. I can stay for the year as long as you pass the spot checks. Carry on your usual schedule during his visits."

Emily said, "What will you be doing?"

"Building our own clone of Hoyer's saucer for travel into the future. The time travel agency can spend the rest of our lives looking for us in the past. We'll be living an idyllic life in the future of our choice."

"What's the second part of your plan?" She put her arms around him.

"To avenge mom, Ken and David. Hoyer kills with impunity outside his time. He'll go right on until somebody stops him."

"How?"

"I have to kill him," Greg whispered.

"Oh Greg ... ."

Greg changed the subject. "We have two things to do immediately. Get a 200 KV transmission electron microscope on order from Hitachi and rent an isolated facility suitable to build our machine in. The problem is we can't do either. Any transaction with our names will come to Webster's attention."

"How will we do it?"

"We need a confederate that Webster and Hoyer are unaware of. Someone we can trust."

"I have several teacher friends, but they don't fit the bill."

"Ralph Gould could do it."

"The young man who killed Antonio and married your girlfriend?

"He took my place in the Romo drag race. As boys, I could always count on him."

"I'm interested in how you're going to enlist him," Emily commented. "Let's see if I can find him in the telephone book." Emily removed a telephone directory from a kitchen cabinet. "Here it is, Ralph and Janet Gould on Strong Street."

"You call, Emily. It could get pretty confusing if one of them recognize my voice. Ralph will remember you. Find some pretense to see him without Janet present."

Emily dialed. "Hello, Mrs. Gould. This is Emily Johnson from the Junior College. Is Ralph home?"

Greg watched her nod affirmatively several times and then hang up. "What did she say?"

"Ralph's left for work at White's Swimming Pool Company. He's a swimming pool salesman."

"Find the address, and let's go."

Twenty minutes later, Emily pulled up to White's Swimming Pools, a one story white stucco building set back from a gravel parking area.

Greg recognized Ralph as soon as he opened the door: still lanky, but his hairline had receded. He was dressed in a polyester leisure suit of the '60s. Greg calculated he would be 33.

"Mr. Gould?"

"Yep, can I help ya?" Ralph asked. His face became puzzled as he stared at Greg.

Greg said, "Can we talk privately?"

"Sure, come on in."

Ralph closed the door and poured them coffee from a brewed pot on a low rosewood table. "You know me; who are you?"

"This is Emily Johnson."

Ralph said, "Hi, Miss Johnson, I remember you — but who are you, mister?"

"Do you remember practicing jujitsu with Greg Philips by the hour when you were kids?"

"Course I do. You must be a friend of Greg's. I understand he's some kinda scientist up north. My wife stayed in touch with his mother until she died of cancer a while back. I ain't seen or talked to Greg in 15 years."

Greg wondered if Hoyer had anything to do with the cancer. "Do you remember riding with Greg when he clipped a man on Main Street and ended up hitting the curb and blowing a tire?" he continued.

"I sure do! Some weirdo in a Buck Roger's suit. He broke my nose!" Ralph looked at Greg. "You know stuff only Greg would and you look a helluva lot like him — 'cept for the age."

Greg analyzed Ralph's friendly reaction to his anecdotes and decided to drop his bomb. "I am Greg. I am fifteen years older in 1968 than I should be because I just traveled back in time from 1983."

Ralph's shoulders stiffened. He sat the coffee cup down and ran fingers through his hair. "You've got some bullshit goin' on here."

"I can prove it. Ask me about anything you and I ever did together, and I'll tell you."

"What's your game, buster?" Ralph pushed himself away from the conference table. Greg recognized Ralph bracing for a fight.

"Ralph, I don't have any game. How about my high school parking lot fight with a Pachuko named Antonio Romo?"

Ralph pushed his chair over. "Give me the truth right now, buddy, or we're gonna go toe-to-toe."

Emily said, "I didn't believe him either until he proved it to me the first time he came here in 1953. You saw him then. Greg and I were having lunch at Nicolino's the day before your drag race. You came in with Janet Ward and sat in the adjacent booth. Greg overheard your conversation with Janet."

Ralph's face was beaded in sweat. "I did take Janet there on our first date. You were there with an old guy who jumped up and glared at me so upset, I thought he was gonna jump my bones. Janet and I wondered about it later, 'cause he'd been the spitting image of an old Greg." As Ralph's memory neurons kicked "You've aged, but he hasn't! What the hell's going on?"

Greg said, "I've been in 1953, 1983 and now 1968 all in a matter of a couple of weeks of my time."

Ralph said, "I can't believe this! I... ."

Emily reached into her purse. "I wouldn't believe it either until Greg gave me this." She handed Ralph the 1980 penny.

For the rest of the morning, Ralph quizzed Greg on minor adventures they'd had together. By late morning, he walked around the table and hugged Greg. There wasn't a dry eye in the room.

Later, over lunch, Greg filled Ralph in. "In the original 1953, I was the one who raced Romo and ended up killing him. Time takes peculiar twists when someone travels in it. The actions of the traveler can have an effect referred to as a 'time split.' One time goes on as if the traveler isn't there, but the other time is altered by his presence. Somehow, time minimizes the alteration and allows the more robust of the two times to carry on. That's what happened with us. There was no real change in the outcome. You raced as I had. My return to 1953 from 1983 changed very little. Romo was still dead, only you got stuck with the guilt, not me. And for that, I apologize."

Ralph sat back, visibly relieved. "Janet will never believe this!"

"You can't tell her. I'll tell you why."

Thirty minutes later, Ralph said, "I get the picture. What can I do to help?"

"I need you to be my eyes and ears. To lease an obscure laboratory for me; to buy me an electron microscope; to get me tools and parts; do my shopping and pay my bills. Basically, be my lifeline to the outside world, while I build the machine."

"I'm just a swimming pool salesman."

Greg said, "Emily and I are rich. Can I count on you?"

"Did'ja ever have any doubt?"

The next few weeks included myriad activities for all three of them. Greg moved into the Mission Inn under an assumed name and had Ralph rent an office on the ground floor complete with telephone, rented furniture, and a company name on the door — Materials Analysis. He had Ralph obtain a business license.

Emily sold enough stock to raise several hundred thousand dollars and took Ralph to Citizens Bank, where she transferred the money into a corporate account for Materials Analysis with Ralph authorized to sign the checks.

Greg spent many hours over several days tutoring Ralph in the basic terminology of electron microscopy. When Ralph was comfortable that he had enough superficial knowledge, he placed a telephone call to Dr. Robert Buchanan. Greg stood beside him with a pencil poised above hotel room stationary. He shared the telephone's ear-piece.

"Dr. Buchanan, my name is Ralph Gould. I'm business manager for a start-up electron microscope service laboratory here in Riverside. I understand you're the general manager in this country for Hitachi?"

"Yes I am. What can I do for you?"

"I've read that Hitachi has recently introduced a 200 KV electron microscope, and we want one."

"I'll be happy to have our sales rep call on you."

"That won't be necessary. My principals know exactly what they want. We want to get a 200 KV EM as soon as possible."

"I'm pleased to hear that, but you must know we just introduced the HU-200A and deliveries are going to be 10 to 12 months."

Greg shook his head vigorously at Ralph. Ralph took his queue.

"That won't do, Dr. Buchanan. We need immediate delivery."

"Well, I am sorry, but there are just none available any sooner. Maybe we can work out a rental on an HU-11A for you until the HU-200A is available. It's a 100 KV system."

Greg shook his head again and scratched a note furiously for Ralph to read. Ralph nodded. "No, Dr. Buchanan. We have to have 200 KV capability. We expect several lucrative contracts from local companies like Kaiser Steel and Rohr for metallurgical applications requiring the penetrating power of 200 kilo volts."

"I understand, Mr. Gould, but even if I put you at the top of the list, the best factory delivery I could get you is nine months."

Greg scribbled again. Ralph read and retorted to Buchanan, "You have a demonstration model in Mountain View, don't you? Sell me that one."

"Well, yes, we do, but we need it for demonstrations."

More scribbling. "What's the price of the HU-200A?"

"$125,000.00."

"We'll give you $150,000."

"I couldn't even take $200,000."

Ralph's countenance was imperturbable. "I'll write you a check today for $250,000."

"Where do you want it shipped?"

After Ralph hung up the phone, they laughed and clapped each other on the back.

Ralph visited several realtors as the representative for Materials Analysis. He found a small yellow, two-bedroom farmhand cottage with lathe board exterior set several hundred yards back in an orange grove. It was only accessible from a rutted dirt path meandering through the grove and not visible from the frontage road. The interior was warmly covered in thick knotty pine paneling, circa 1920. An ancient, long-retired water tower set atop a large windowless square grove storage room was stationed several yards into the grove. Ralph drove Greg to see it in his dusty Ford pick-up.

Greg said, "It's perfect. We'll clear out the storage room and use it as the lab. Lease the place immediately and then get an electrician in. We're going to need electricity in the storage shed."

Emily shopped for furnishings for their new home. The day the furniture was delivered they invited Ralph to join them for dinner.

"Janet's getting upset with me being gone all the time. She's thinking I'm having an affair."

Greg frowned. "You better get back to your old routine for awhile. I'll meet the electrician. When will Hitachi be ready to start the installation?"

"Tuesday of next week. Buchanan's sending their best local engineer, a Mr. Naito, to do the installation."

Greg said, "Well, unfortunately you'll have to be here for that. Naito knows me from SRI in the '60s."

The next morning, the electrician arrived; he was a slight man with dark hair growing back from a wide forehead. His face was immediately familiar. Pete Leeper, late of the old Strong Street gang. As Leeper ambled into the room with his familiar bustling gait, Greg saw his face reflecting recognition.

Greg said, "We need to install some electronic equipment in the old water tower out back. We'll need a hundred amps of 220, and 200 amps more of 110 by next Tuesday. Is that a problem Mr.—-?"

"Leeper, Pete Leeper. Sure, I'll have it in by tomorrow night. We need a building inspector to approve it before we can turn it on. I'll schedule a Monday inspection. By the way, are you a member of the Philip's family?"

"No, my name is David Hoard. Nice to meet you, Mr. Leeper." He extended his hand. Leeper said, "You sure look like an older version of an old buddy of mine."

"Sorry, I can't help you." Greg shook his head.

That evening, Greg told Emily and Ralph about the incident.

"I've got to be a lot more careful. I think I'll grow a beard."

The weeks passed. Mr. Naito finished the installation while Greg spent his days with slide rule in hand performing calculations. He made copious lists of things for Ralph to buy. They kept an accounting of the calendar and the days they knew there would be a saucer sighting from Miss Hernandez research. On those days, Emily would return to her duplex and single lifestyle, including the walk home from school.

In early April, Greg telephoned Ralph at White's and asked him to swing by after work. "Ralph, I need you to fly up north and visit a company on my behalf."

"Sure, what's up?"

"I need a custom-designed high vacuum system to interface with the EM. I'm sending you to a Perkin Elmer division in Mountain View. I know them from work they did for us at SRI."

"What do I have to do?"

"Deliver my design drawings and emphasize getting the fastest delivery. I want you to call the president, Bob Ward, and make an appointment."

Ralph said, "I won't be able to explain your drawings."

"My design is pretty complete. If they have any questions, call me and I'll talk to them. Your purpose in going is to make it clear we'll pay a premium to be at the top of their production schedule." Ralph was put through to Bob Ward and made his appointment for the following Tuesday.

Tuesday morning, Greg was sitting at the EM adjusting parameters to set a mini-black hole in motion. The telephone was by his side prepared for any calls from Perkin Elmer. Greg heard pounding at the front door and decided to ignore it. The pounding increased. He opened the door. Janet Gould stood before him, her face flushed in anger. She was dressed casually in loafers, slacks and linen shirt. She had retained the winsome figure he remembered.

Greg said, "Can I help you?"

"I'm looking for my husband, Ralph Gould. He's here, or will be shortly."

"There's no one here by that name."

"Baloney. I've found this address on several receipts at home, and in my husband's little black book." Janet side-stepped past him into the living room and looked around. "Is your wife here? I suspect she'll know where he is."

"I don't have a wife. There's no woman here."

"Oh yeah, then I suppose that woman's rain coat hanging on the coat rack is yours." Her lower lip started to quiver.

"Dammit, help me. I know my husband is having an affair, and everything points to this address. He left last night without telling me where he was going."

Greg felt trapped. "I know your husband, but he's not having an affair. He's just working for me."

Janet wasn't listening. She was studying him with fire in her eyes. "That's you under that beard, isn't it, Greg?" She whispered. Greg inhaled and almost hyperventilated. "What?"

Janet walked across the room, then turned back to him, took his right hand in hers and turned it over. "There's the scar you got breaking up a dog fight when you were eight. Don't you think I'd know you anywhere? I remember every scar, birth mark and mole on your body. There wasn't much we didn't know about each other."

Greg sat down wearily on a naugahyde sofa Emily had recently purchased.

Janet said, "What's going on? What are you and Ralph doing? Why are you lying?"

Greg thought through his answer. "I'm sorry how this sounds and looks, but Ralph isn't having an affair. His secretive behavior has been to help me." Greg took a deep breath. "He's out of town until Thursday doing business for me."

"Why won't you admit you're Greg? Where did Ralph go?" Janet's eyes were ice blue and suspicious. "Give me straight answers, or I'm going to the police."

"Yes ... I am Greg and Ralph's helping me get an electron microscope service laboratory started." Greg's composure was slipping. "I'm trying to keep a low profile for personal reasons."

Janet looked at him intently for several moments and then moved to the sofa and sat beside him. "Your mother told me you were a scientist, so that makes sense." She took his hands in hers and looked into his eyes. "She also told me you married a spoiled southern belle who was making your life a living hell. You're running from your wife, aren't you?"

Greg stood and grasped at her conclusion. "I want her to think I'm dead or something and give up on me. She threatened to ruin me if I ever left her. Ralph is helping me get a new start."

Janet got up from the sofa; her suspicions somewhat assuaged and commented, "If that's true, you must have had a helluva life. I'm sorry to tell ya, but you look years older than you are."

Her eyes faded into the past. "Why did you leave me?"

"I'm truly sorry." Greg put both hands on her shoulders.

"I worried about your long-term commitment." She reached out and squeezed one of his hands. "Ralph didn't have your ambition, but he was content to get a job and settle down, but you were always talking about going off and becoming somebody."

Greg withdrew his hand from hers and turned his head toward the kitchen. "Can I get you a drink? Coke? Beer?"

Janet laughed, "Yeah, I'd like a beer." She walked beside him into the kitchen while he opened two bottles.

"I visited your mother regularly and stayed current on you. We couldn't understand why you avoided Riverside."

"When did you find out mother had cancer?"

"That was the strangest thing." Janet's smile faded. "She'd had a physical a few weeks before and got a clean bill of health. Then, on a visit a little later, I was stunned. She looked ghastly. She told me that almost overnight her strength was gone. She could hardly get around. She went back to the doctor. She had advanced stomach cancer." Janet took a large swig of beer. "I visited her after the operation, and she confided she was dying. I wanted to call you. She said no; you had enough to worry about."

"I should have been there."

Janet took another drink. "What exactly are you and Ralph doing here?"

"I'm starting an electron microscope service laboratory for the big metallurgical companies in the area. I've got to make a living. Ralph is my part-time business manager and interfaces with the world. I have to stay undercover until Helen stops looking for me."

"You poor baby." Janet set her beer down and put her arms around him and pressed her cheek into his chest. "You've really had a rough time of it. I feel like it was yesterday." She giggled, "I feel like it's Friday evening, and we're heading out to the Rubidoux Drive-in movie."

"That was all a long time ago."

She set her bottle down, raised her face, and kissed him passionately. She leaned back and said, "He wasn't reluctant to chase me in those days, and I'm pretty sure he's not being faithful to me now. Maybe we should pull the drapes and pretend we're back in that old Ford of yours'."

Greg pushed her away and said, "Ralph is my friend and so are you. We have wonderful memories of our youth together. Let's leave it at that."

Janet looked as if she'd been slapped in the face. She composed herself and said, "I don't know what got into me. I'm sorry. I'd better go."

Greg put his arm around her shoulder and walked her to the door. "It's been wonderful seeing you again. When Ralph calls, I'll tell him to telephone you immediately."

Janet scurried to her Toyota and looked back at him as she opened the door. Her moist eyes shone with humiliation. The phone was ringing as he shut the door.

"Hello, Ralph. How's it going?"

"It's a done deal. But it's gonna cost ya. Since your design is so unique, they wanted time and materials, plus a fixed profit percentage on the total. I couldn't blame them."

"Delivery?"

"Bob Ward agreed to squeeze us in for a September delivery. I offered a 25 percent bonus for August, and we struck a deal."

"Great, Ralph. Greg paused. "Janet showed up today."

"Oh, no! What happened?"

Greg gave him an edited version and added, "I told her you'd call her."

Chapter 20

While Greg awaited the fabrication of the machine's bell jar, he duplicated the forward time travel parameters he'd observed in Hoyer's saucer. He and Emily's evenings were quiet and intimate. Greg relished the restorative idleness and solitude. The occasional oddity of perception when he would unexpectedly see Emily anew, quietly preparing dinner or brushing her hair.

Everything about her was fresh and exciting and all he'd ever wanted in a mate. He admired the undulant and elastic movement of her body, the curiosity of her mind about his project and realized her sexiness was a harmony, an orchestration of all her attributes.

Shortly after Janet's visit, they invited the Goulds for dinner. Janet had remained silent through most of the meal, but after more than a little wine asked, "Weren't you Greg's teacher in high school? Don't you think you're a little old for him?"

Ralph said, "What's with you? Ease off the wine."

Perkin Elmer delivered the cylindrically-shaped bell jar in August and, with the help of a rented fork-lift, Greg and Ralph were able to wrestle it into position. Once squeezed inside the storage room's old wooden doors, it was awesome — seven feet tall with a ten foot diameter and rounded external surface. Greg's design included a provision to drop the EM down through a center orifice with a chain hoist and bolt it in place like the hub of a wheel.

Greg spent the next few days bolting everything together. His mind kept returning to his conversation with Janet about his mother's sudden onset of cancer. He broached the subject with Emily over dinner.

"I always thought my mother was just unfortunate in contracting stomach cancer. When I visited her in the hospital in 1966, she didn't mention anything about a clean bill of health just a few weeks before."

"Do you think Hoyer went into her past and caused the cancer the way he did her polio?" Emily offered a bowl of mashed potatoes. "I shudder when I remember his threat to cause her a new misery every day until you capitulated."

Greg shook his head. "He doesn't make risky and unnecessary time splits unless they help achieve his ends."

"Maybe he has let you know somehow." Emily picked up their plates and walked to the sink. "Maybe he exposed her to cancer, and then waited for you to come forward."

"How did he let me know?"

"Maybe a personal message in newspapers down through the years. He knows you communicated that way with Ken across decades."

Greg considered Emily's comments. "It's worth a try at finding out."

"I'll drop you off at the library on the way to school in the morning. If he did try to tell you, you might as well know it."

The next morning, Greg requested copies of the local papers for the months of October through December of 1965, and January through June of 1966 from a tall gray haired woman in her 60s. He spread them out on a large library reading table and spent the morning reading through them. He looked up once briefly and noticed the librarian staring at him. A wave of apprehension swept over him. He found Hoyer's message in a Sunday edition dated March 13, 1966.

Greg, if you're reading this, then you know your mother is dying of cancer, and we've found you. I warned you about her miseries. Contact me immediately through this paper if you want to spare her. See you soon, tough guy.

Herman

Greg felt sick as he returned the newspapers to the librarian and asked, "Do you get many requests for these old issues?"

"Why, yes. A nice young man has stopped in and asked for the previous four weeks' issues every month for the last two years. He was in just two days ago. He was looking for someone in particular. You'd more or less fit the description he gave if you didn't have a beard. If you're looking for him, I can pass on a message."

Greg stood raised his arms overhead and stretched and twisted. "I don't know anything about him. I asked because ... I'm part of a group that plays Dungeons and Dragons once a month ... I thought one of the clues I'm looking for might be in one of these issues. I wondered if any of my fellow players came to the same conclusion."

He waited in the dim interior of the Mission Inn bar nursing a beer until Emily arrived.

"I screwed up; you were right. I found a message to me from Hoyer in a Sunday issue in March 1966."

"Buy me a drink."

Greg waved for the barmaid. Emily asked, "How did you screw up?"

"The librarian said a young man has been coming in looking for a response to this personal every month for two years. He gave her my general description. I denied it could be me."

"Oh, God!" Emily exclaimed. "Webster will find out you're here on his next visit."

"He was here two days ago. We've got a month."

"Can you finish the saucer in that time?"

"I'll try."

In the days that followed, Greg worked a routine of eight hours on the job followed by two hours of deep sleep. One Wednesday, Ralph stopped by. "Greg, I don't know how to say this ... Janet mentioned during dinner last night that a young man stopped by our house several months ago and asked if she'd seen ya." Greg laid down a screwdriver and turned.

Ralph continued, "She told him not in fifteen years. She didn't think much of it at the time, but now she's added it to the mystery of ya. She doesn't buy your story. She refers to your 'hidden agenda' and 'dark past.' Can ya finish and be out outta here before he comes?"

Greg looked at the floor and pondered for a couple of minutes "No. You'd better take Janet out of town on the day we know Webster's returning. I'll sponsor a vacation for the two of you to someplace ... to Hawaii. Okay with you?"

"Sure! But if he really digs, he may find I've leased this place and wonder why."

"I'd better be gone too. Hell, if he makes it this far, he'll go over this place with a fine tooth comb. He'll see our tracks coming out here from the house and find the machine."

"Don't panic, Greg, we got three weeks to let the friggin' weeds grow if we stop walking across 'em." Ralph insisted. "The way weeds grow around here, with a little sprinkling we can have them tracks covered."

"You're right. We can go out the front door of the house and circle around through the orange grove to the rear door of the shed." Greg paused. "We'd better authenticate the Materials Analysis office, as well. It needs an electron microscope in place,

"Another $100,000? How rich are you anyhow?"

Greg said, "We'll buy a used one. Call the RCA Electron Microscope Service Company in Los Angeles. RCA used to be the biggest thing in EMs until they decided to move their EM design engineers into color TV design. In 1968, they're trying to dump an inventory of used model EMU-4s. You can pick one up for a song. Have it delivered immediately."

Emily was wringing her hands. "You don't expect to make it?"

"I'm going to try like hell, honey, but Murphy's Law is always in effect. Tell the school you're leaving in three weeks for an indefinite period due to whatever excuse you think might fly."

Over the next two weeks. Greg installed a bowling ball and put settings into his machine he'd calculated based on parameters he'd observed in Hoyer's saucer. The ball was swallowed by the resulting black hole but didn't reemerge.

Disappointed, Greg started varying one parameter at a time and observing the results. The process was slow and frustrating. Two days before Webster was due, he shut the machine down. "I thought I knew the secret to Hoyer's saucer's forward time travel — but I guess I don't."

Ralph was nursing a beer. "What are ya gonna do?"

Greg pushed back in his chair. "Close up the place and put up a for rent sign. We have to assume Webster will learn we're in 1968. Ralph, before you catch the plane to Honolulu tomorrow, take $50,000 out of the joint account. Keep $5,000, and give the rest to Emily in one hundred dollar bills or smaller. Emily, cancel all utilities and get a post office box for your mail."

Ralph set his beer down and stood up.

Greg added, "Ralph, please get Emily and me train reservations in the names of ... Mr. and Mrs. ... Ronald McDuffie out of Los Angeles to ... where would you like to go for a hideaway for a few weeks Emily?"

"I've always wanted to see New Orleans."

"New Orleans it is. Ralph. What hotel do you have reservations at in Hawaii?"

"The Reef. We've been there before."

"I'll call you there when we get to New Orleans. I'll say it's ... Mr. White from your office. We've got to have to plan each step as we go along — based on what Webster does."

The following Monday afternoon, Greg and Emily caught a Greyhound to the Los Angeles massive, salmon-colored Union Train Station. They paid cash at the Railpax counter for a sleeper on the Sunset Limited scheduled to depart in an hour.

That evening, relaxed in the dining car, Emily said, "You told me you were trying to prove how out-of-body spying was possible when you suddenly realized you'd made a black hole suitable for time travel. Do you think you might have proved your out-of-body stuff if you hadn't bumped into the black hole?"

Greg smiled, "Yes, eventually. Urie Geller's ability to describe sites thousands of miles away couldn't have been faked. He had to have been in two places at once to do it."

Emily nodded and then asked quietly, "What do you think is the worst thing that can happen to us now?"

"Webster finds our machine and reports it to Hoyer, and Hoyer comes in for the kill."

Emily shuddered. "Is there a best scenario?"

"Sure. Webster arrived in Riverside today, and the librarian was out sick or didn't tell him about me, or she told him the Dungeons and Dragons story and he bought it. We'll go home, figure out future travel for our machine and be on our way."

"We haven't talked a lot about what will actually happen to us when we travel into the future."

Greg chose his words carefully. "If my calculations are off the mark, the wormhole tunnel might send us into a random time that could be devastating. When the time comes to go, I'll understand if you want to stay behind."

Emily said, "You do have a way of changing blue skies to gray." Greg smiled, "We'll get into the future. I just don't know how yet."

"Greg, you said you were able to look at everything inside Hoyer's saucer. Was there anything you saw that you didn't recognize? Something you might be overlooking?"

Greg thought for a moment. "The monitor."

"What about it?"

"Well, I believe I understood the input data Hoyer punched into the computer. I think it was the date and hour he wanted to travel to plus a longitude and latitude. What I couldn't figure out was a slender rectangular bar across the top of the screen."

Emily asked, "What did it look like?"

"It was equally divided up into five parts." Greg picked up his paper napkin and sketched the rectangle. "From left to right the first three squares were colored green. The fourth colored orange, and the last one to the right was flashing red. Oh, and it had overlaid numbers 42873745, and the orange had overlaid 42873845." He pointed his pen at the squares.

"Mmm, very similar numbers. Any others?"

"The first green square had 42864545 overlaid and ... about the first quarter of the square was pulsating a darker green than the rest."

Emily thought for several minutes while Greg cleaned his plate. She said, "It sounds like some kind of cryptogram."

"Cryptogram?" Greg set his fork down. "Do you know something about cryptograms?"

Emily smiled. "I do have some hobbies you know. Studying cryptography is fascinating. The first people who used codes successfully were the Spartans 2,500 years ago — a coded papyrus wrapped around the general's staff. In our country, Thomas Jefferson invented code wheels. For centuries, armies have tried to decode messages from their enemies. Japan sent a coded message to their Washington D.C. Embassy a few hours before they attacked Pearl Harbor. Our ciphers decoded it but were unable to get the message to their military commanders in time."

Greg shook his head. "You never stop impressing me."

Emily continued. "There are many code types: tic-tac-toe codes, monoalphabetic codes, substitute codes — many, many."

She picked up Greg's napkin and stuffed it into her purse. "Let me see if this is some sort of cryptogram."

Chapter 21

NTSA - 2030

Astronaut Cody Webster landed his Mercury Four in the tight security of Goddard Space and Time Center outside Washington D.C. and caught a taxi directly to Hoyer's office. Hoyer's receptionist told him, "The director wants your trip report before seeing you."

Webster removed a quarter sized-disk from a palm size computer in a vest pocket of his uniform and handed it to her. She fed it into a slot in her desk. Webster moved his lanky frame to a visitor's chair, his manner quiet and noble with no evidence of the antipathy he felt for the man inside.

After a few minutes, the brunette nodded for him to go into the director's office. Webster stood at attention within the great man's bastion. Hoyer sat before him, his eyes closed as he inhaled deeply on a cigarette.

Hoyer opened his eyes and observed Webster, whose pinkish skin appeared weather-cured from a recent sunburn. "You seem to have returned from a rather inconclusive journey. Tell me about it."

Webster shoved one hand deep into a Velcro-lined pocket of his uniform trousers and gestured toward the desk with his other hand. "As my report indicates, I'm focusing in the three locales Philips was familiar with; the San Francisco Bay area, Savannah, Georgia — where I got this sunburn, incidentally — and, to a lesser extent, Riverside, California, where you found him the first time in 1953. He may have gone back into some other time, but I doubt it."

Hoyer said, "That's not the message he left Southwell before traveling this time."

"We don't think he had any money or identification, let alone proper immunization for an earlier time. My conjecture is he's in a time and place where he can easily establish himself and make a living."

Hoyer raised his voice several decibels. "Skip your conjecture, and tell me about the Dungeons and Dragons guy."

"The man the librarian described in the 1968 visit could have been him, although she couldn't describe his face because of a full beard. I'd never heard of Dungeon and Dragons, D&D, but soon found out it was a rage in 1968, and the story of looking for D&D clues was not unreasonable. I went directly to the Junior College to check out Emily Johnson and was told she'd applied for a leave of absence several weeks ago to go east and care for an ailing relative."

Hoyer leaned forward. "Where in the east?"

"The girl in the office didn't know. Then I tried looking up an old girlfriend of Philips."

Hoyer said, "What old girlfriend?"

"Her name now is Janet Gould. I first came across her when I was scanning Riverside through the 1950s. I looked up the young Greg Philips and found him with her much of the time." Hoyer said, "Get to the point."

"I visited her parents two months before the D&D guy hit the library, and they told me she was married to a Ralph Gould and gave me her telephone number and address. I visited her and struck out. She hadn't seen or heard from Philips in years. After hearing about the D&D, I tried to visit her again, but her neighbors said she and her husband were on vacation."

Hoyer frowned. "It's too pat. Somebody who looks like Philips shows up and both girlfriends leave town — too much of a coincidence.

I'm going to assign two more time astronauts to assist you. Go back to 1968 and dig deeper. Check out Emily's neighbors, Janet Gould's activities and neighbors. Check out her husband. Find out when he came into the picture. Stay with it until you find Philips."

Chapter 22

Greg and Emily were escorted to their table by an elderly waiter. The restaurant was Antoine's, one of New Orleans' finest. The wall decor included numerous old newspaper clippings, photographs, and previous years' menus. The waiter strolled to a table adorned with snow white napery and gleaming goblets and spread napkins over their laps with shaky hands.

It was Wednesday evening. They had arrived to a radiant New Orleans earlier that morning and taken a taxi directly to the Vieux Carré, the original New Orleans, popularly referred to as the French Quarter. They had checked into the Weston, a stone's throw from the Quarter and a couple of blocks from the Riverwalk Marketplace. After checking in, they cleaned up and dressed in their lightest clothing, and did a walking tour of the old town, wandering through the quaint historic Garden Area, admiring old Gothic cottages with heart pine floors, and free-standing staircases and Doric columns. They walked through St. Patrick's cemetery, where the dead were buried above ground in magnificent stone tombs. They listened to the melodious accents of passing Cajuns and ended up in front of Antoine's restaurant.

The waiter placed a beer and 7UP within reach. They silently toasted each other. Emily, with a twinkle in her eye said, "I figured it out."

Greg splashed a little beer onto the tablecloth. Emily pulled a folded sketch out of her purse and spread it out. "Well, I tried my hand as a cryptanalyst. At first, I went through several of the more esoteric decoding types. I finally settled on Julius Caesar's."

"Caesar's?"

"When he wasn't busy running the Roman Empire, he enjoyed

dabbling in code making. One of his favorites is still around. It's called Caesar's Code and part of it is to replace letters in the alphabet with letters standing three places further down the line. I used that concept and see what happened to the numbers in the far left square." Emily pointed and said, "42864545 becomes 19531212. The numbers in the orange become 19540312 and in the red, 19540412."

"So?"

"So, 1954 stands for the year, the next 12, for December, and the last 12, for the day of the month, the day I think Hoyer arrived in Riverside."

"Why the other dates?"

"The orange is March 12, 1954 - three months later and the flashing red is April 12th, exactly four months to the day after he arrived."

Greg leaned forward to study Emily's sketch. "Then, this is a constant reminder of just how long he could stay during that trip."

Emily nodded. "Remember he said Congress had placed very strict guidelines on their travel."

"My God, Emily, you're right!" Greg pointed at the first two colored green squares. "What do you make of this?"

"The green square is the month of December, and the pulsating dark green is how long he'd been traveling at the point he took you back to Menlo Park."

Greg sat back. "But why the code?"

Emily answered, "I've given that a lot of thought. I think it was just a simple code in the event the saucer crashed or something and people in local time found it."

Greg nodded. "Our ancestors don't want people in local time to know they were entertaining time travelers ... Emily, you did great."

"Thanks." Emily smiled. "What happens now?"

"We need to know if we're in danger. I'll call Ralph in Honolulu in the morning and ask him to call here as soon as he gets back to

Riverside."

"Are you close to a solution for your saucer problem?"

"Not yet."

"When you find it, where are we going?"

"First stop will be 2030. I've got to keep my promise to Ken. After that, I think we'll play it by ear, depending on how we like the 21st century."

After dinner, a light drizzle started and amoeboid pools of water dotted the cobblestones. They walked down St. Peters Street to Preservation Hall and spent an hour listening to the ancient jazz musicians before calling it a night.

A hazy sun came up over the west bank of the Mississippi, penetrated the stormy clouds and invaded their room. Greg placed the call to the Reef Hotel in Honolulu. Ralph didn't answer, and the Reef operator reminded Greg it was two in the morning in Hawaii. Greg left a message asking Ralph to call the Weston.

Greg asked Emily, "How about some breakfast down in the Quarter?"

"Sounds great."

They walked out into a light drizzle. The smell of last night's jambalaya, booze and blackened catfish dusted the still morning air. They had a traditional breakfast of fresh roast coffee and beignets in the Café du Monde and then walked it off along the levee, watching commerce floating down to sea and listening to the desolate breath of the old Mississippi. At noon, they reentered the Weston in anticipation of Ralph's call. The phone was ringing as they entered the room.

Greg answered, "Yeah, Ralph?"

"I blew it Greg. Webster got to Janet."

"How in hell...?"

"Yesterday afternoon. We got into a spat. I went to the bar and filled up on Mai Tais. She went to the beach and told me later Webster approached her there. I'd screwed up big time. I used a credit card to buy the tickets and he traced me with it."

"What did she tell him?"

"That she'd seen you in Riverside. You were there starting an electron microscope service business and seemed very mysterious."

"That's all?"

"That's all she knows."

Greg asked, "How did you get this number?"

"It was on a call-slip, shoved under my door when we got in a little after two. I was too bombed to call you. I'm awfully sorry about the screw up."

"What's done is done. Come home on the next plane. We've got to find out if Webster made it to our machine. I'll call you Saturday morning between eight and nine at ... Mike's Cafe on Market. You know it?"

"Sure."

"You'd better swing by Materials Analysis, as well, and see if Webster's been there."

"You betcha. And buddy ... " Ralph lowered his voice, "I won't let ya down again."

Greg tasted a drop of bitterness as he replaced the phone on the receiver.

"What is it?" Emily squeezed her hands.

"Webster knows I'm in 1968.

Chapter 23

Janet was crabby and complained of a headache on the flight to Los Angeles and once home went directly to bed.

Ralph backed out of the driveway. Within a few blocks, he noticed a late model Ford behind him. It followed at a distance until he entered the downtown traffic. Ralph watched the car in his rear view mirror pass a pickup and squeeze in behind him. He thought, "Am I being paranoid, or could that be Webster?" He slammed on the brakes. The Ford swerved to the left to avoid a collision and sailed past. Ralph relaxed when he saw that the driver was a balding middle-aged man.

Once in the Mission Inn complex, Ralph walked down the corridor to the Materials Analysis office and unlocked the door. As he was closing it, it suddenly slammed back open, and Ralph was propelled into the room. When he got his footing, he turned and faced the balding driver pointing a small weapon at his mid-section.

"You must be Hoyer." Ralph slammed the chair forward knocking the gun out of Hoyer's hand. He kicked the chair aside and swung a cocked fist at Hoyer. Hoyer caught Ralph's wrist and spun him around, grabbed Ralph's neck and jerked him off his feet. Ralph went limp. And then, taking advantage of the momentum, leveraged his body to pitch Hoyer over his shoulder. Hoyer recognized the move and slammed a fist into the side of Ralph's head. Ralph lost consciousness.

Chapter 24

NEW ORLEANS - 1968

E mily exhaled, "Ohmigod, can he find us?"
Greg shook his head. "Only Ralph knows where we are. As far as Webster is concerned, we could be anyplace on the face of the globe. We'll stay holed up until Ralph finds out if Webster has found our machine."

Greg spent the afternoon on his time travel problem. After a night of tossing and turning, he said in the morning, "Maybe we should get another hotel as a precaution."

Emily nodded and started packing. Minutes later, they checked their bags with the bellhop and paid the bill. Greg went to the bar and called the nearby Marbury hotel and made reservations. As he came out of the bar he froze. A tall man in his late 20s was standing behind a column and intently looking at Emily sitting in the lobby. He had a full cherubic face at odds with his narrow-shouldered physique. His hair was bushy with sideburns down below his ears.

Greg ducked back into the bar and dialed the front desk. The desk clerk listened and walked across the lobby and whispered to Emily. She picked up the house phone.

"Don't look around. You've got a tail."

"What shall I do?" Emily was surprisingly composed.

"Go to the ladies room and stay there until you hear a commotion. Then hurry out the back way and get to Jackson Square. I'll meet you there when I can."

Greg peeked over the swinging bar door and saw the tall man

stiffen as Emily put down the phone and walked briskly into the ladies' room. He looked unsure if he should follow her. Greg took that opportunity to trot softly across the room and approach from behind. At the last instant, the tall man heard his approach and spun around. Too late. Greg dropped him with a single karate chop to the base of his neck, turned and ran out the front entrance. The clerk, bell boys and several guests stood immobilized, stunned by the silent violence they had just witnessed.

On the street, Greg jumped into the first cab in line and gave instructions. As the cab accelerated, Greg panned the front of the hotel through the cab's rear window. Another lanky young man with a weathered sunburn was staring at him with a shocked expression. He grabbed the next cab in line.

Greg's cab took him to the front entrance of Rubinstein Brothers on Canal Street. Greg threw money at the driver and dove out the door as the following cab screeched to a halt.

Greg ran into Rubinstein's and out the rear onto Magazine Street. He jumped in front of traffic causing a cacophony of horns and squealing brakes.

He ran full out down Royal Street and its rows of antique shops. He darted into one of the shops, and sped through and out the rear door. As he rounded the next corner, he glanced back and saw the sunburned young man exiting the antique shop and gaining on him. Greg ran through seedy bars on both sides of Bienville, a street that was a bouillabaisse of squalor. He selected a bar he could see straight through to the alley behind and raced into it. He leaped over the backdoor steps and slipped on the alley's greasy surface. He hit hard on his twisted buttocks and a bolt of pain shot up his lower back. He hobbled out to the street and hailed a passing cab. The sunburned man skidded to a halt as Greg's cab pulled away, and he frantically began ferreting out cabs disgorging people.

Greg took the cab to Riverwalk, a quarter mile packed with shops, cafés and restaurants on several levels. His back pain had subsided as he stepped out and paid the cabbie. He entered the first men's clothing store and purchased an undistinguished full rain coat and hat. He put them on and was satisfied he was partly disguised.

He strolled onto Poydras Street and hailed another cab. He asked to be slowly driven into the center of the Quarter. He saw Emily trying to look inconspicuous in the center of Jackson Square. He had the cab stop just long enough to painfully step out and call her.

Emily whispered, "I saw people around the young man lying in the lobby. Did you kill him?"

"No, of course not. He was just unconscious. They're here, honey. When I got outside, a guy that fits your description of Webster was there. He just chased me all over town."

"Where do ya wanna go?" The black cabbie asked.

"Shreveport."

"Shreveport? Man, you crazy? I'm a New Awlens cab, that'd cost you a fortune."

"Just start the meter. When we get there I'll pay you double."

"Man, in New Awleans every man's a king and every woman's a queen!" His rosebud mouth split a shiny, ebony heart-shaped face in two with a handsome grin. He turned the cab around illegally on Erato Street as they were approaching the Greater New Orleans Mississippi River Bridge.

The astronauts searched New Orleans for the rest of the day. Webster and Francisco canvassed the city while Ellis checked the airport, train station, and bus depot. They scoured the French Quarter's farmer's market, every oyster bar and sidewalk café. They met at the fountain in front of the Café du Monde at the end of the day, perplexed.

After two more days, they located the cabbie who had driven Greg to Shreveport. They moved their saucer to a secluded bayou outside Shreveport and sent it to storage.

Francisco and Ellis split-up in pursuit of Greg and Emily. Commander Webster stayed next to a Shreveport motel room telephone in a hastily arranged command headquarters.

Bill Ellis at age 23 was the youngest of the three and had been at the top of his class in the Air-Time Academy class in Colorado Springs. He was slender and on the short side for the academy, with a sandy cowlick forever sweeping across his freckled forehead. "I think I've found where they went, Cody. I met a ticket vendor who remembered a good-looking dish-water blond accompanied by a bearded man. He vaguely recalled them catching a bus last Thursday evening headed for somewhere west."

There had been half a dozen buses heading west and Francisco and Ellis divided them up. Francisco found a driver that remembered the couple on his run to Wichita Falls. Additional legwork pointed to Albuquerque. Francisco called Webster and was ordered to search out an isolated patch of desert as a landing site for the Mercury-Four. Webster would join them in Albuquerque in the morning but first had to report in to the director as dictated by the stringent time travel guidelines.

The U.S. House of Representatives had formed the Time-Travel committee to establish these guidelines in 2015. The guidelines were soon referred to as the "bible" by time astronauts. Time travel was in its infancy and the scientific community had a dichotomy of opinions about the future effects of "time splits."

The Los Alamos Nuclear Laboratory was converted in 2015 to study time splits via computer simulation. Trillions of computer time split simulations had been performed, and the data was still being analyzed. Congress slapped severe restraints on the astronaut time-travel until the analysis was completed, and the search for Philips was only reluctantly approved with severe restrictions. Hoyer had made the case that it was more dangerous to leave Philips out of his time causing possible catastrophic time splits, than it was to continue the search.

Webster had received initial approval for a single visit of one hour or less of local time, in each month from December, 1953, forward until Philips was located. Once Webster reported Philips was in 1968, the Mercury-Four command computer was programmed to allow travel only to and from 2030 and 1968. They could use Mercury-Four as rapid transportation within the approved time. Webster had to report in with each major find.

Chapter 25

NTSA - 2030

C ody Webster stood at attention before the incensed Director.
"You mean to tell me that after I told you the city he was in,
you lost him? You let a man almost 50 years old beat your entire team?" His eyes were flashing in fury.

"Tony, Bill and I thought it would be easy. New Orleans isn't that
big a city in 1968, and we divided up the hotels between the three of
us from the yellow pages. Gave the clerks at each hotel their description. Gave bribes when necessary. The second day, Tony called me
from the Weston. He'd spotted Emily in the lobby. I tried to get hold
of Bill to help, but he was doing the hotels downtown. I figured we
had surprise on our side and there were two of us to handle one old
guy."

Hoyer glared at him.

"Anyhow, Tony didn't hear him coming until it was too late and
Philips took him out. Next thing I knew, I saw Philips running out just
as I arrived. I chased him all over the French Quarter, but he could run
for an old guy. I'm sorry, sir. I know how hard it is to find someone
even when you know the time he's in. After I reported he was working
with Gould in '68, you told me within hours he was in New Orleans.
May I ask how you knew?"

Hoyer ignored him. "Where do you think he's going?"

"Home."

Hoyer grunted, "Go back, and leave this message for him in the
local paper." Hoyer scribbled on a notepad and tore it off.

Chapter 26

Riding north through the emerald green marshes to Shreveport, Emily posed the question to Greg, "How did they find us?" "We're going home to find out. It shouldn't have been so easy. Last time Hoyer was willing to kill Ken to circumvent years of looking."

The first bus out of Shreveport took them into Dallas-Fort Worth, arriving before sunup Friday morning. They changed buses for Wichita Falls and on into Oklahoma City. They made another bus change in Oklahoma City on Friday night for Albuquerque. The Greyhound pulled into the Albuquerque depot at nine in the morning. Emily took Greg's elbow as he hobbled down the bus steps grimacing at the pain in his back. The driver had announced they had a 45 minute layover for breakfast.

Emily ordered breakfast while Greg found a telephone booth and obtained the Mike's Cafe number through information.

"Who did you say?" A female voice asked across the poor connection.

"My name is White, I'm calling for Ralph Gould who's having breakfast right now in your cafe."

"Just a minute."

Greg waited apprehensively for the sound of his old friend's voice. Several minutes seemed to pass before the female voice came back on the line. "Nobody by that name here."

"Are you sure? Could he be in the men's room?"

"Nope, sorry."

Greg returned to Emily and found a plate of bacon and eggs awaiting him. Emily was eating. "Well?"

"He's not there."

"Not there?" She set her fork down.

"I'll call him at work when we reach the next stop in Gallup." He pushed away from the linoleum table covering.

Emily asked, "Aren't you going to eat?"

"I don't think I could hold it down."

Greg called White's Pool Company from Gallup. The phone rang several times before a male voice answered.

"Hello, I'm a friend of Ralph Gould. Could I speak to him."

"Not here. We don't work Saturdays 'cept me. I own the place."

"Mr. White, do you know where I might reach him?"

"As a matter of fact, we're all a little concerned about ol' Ralph. He called from Honolulu and said he'd be in to work yesterday. Never showed. His wife called last night. She hasn't seen hide nor hair of him since they got home Thursday. Well, I expect I'll see him Monday. What's your name, so I can tell him you called?"

Greg quietly hung up the phone. A wave of fear surged over him.

Later, on the bus, he told Emily they'd stay over at the California border in Blythe until they located Ralph. In the wasteland of his sleepless night, Greg silently prayed for Ralph.

They pulled into the Blythe depot Sunday evening. Emily commented that the little farm community of a few thousand hadn't changed since she was a girl. They checked into a motel and slept, without dreaming, until the Colorado Desert sun spilled through the blinds mid-morning. Greg washed his face and, with a shaking hand, went through the motel operator for White's Pools in Riverside.

Ralph's secretary came on the line. "I'm sorry Mr. McDuffie. We're as concerned as you are. It's not like Ralph. His wife called the police this morning."

"What are we going to do?" Emily asked, tightly clutching her right hand with her left.

"I'm going to call another old friend to help."

He went through Riverside information for Pete Leeper. Mrs. Leeper came on the line and told him he could find Pete at his shop. She gave him the number. He was out on a call and wouldn't be back until late afternoon. At four, Greg called again.

"Pete? It's Greg Philips."

"I knew it had to be you out at the Highgrove house. I knew you'd call to explain. How the hell are you, ol' buddy?"

"Pete, I need your help."

"Why, sure, but first tell me what you've been doing' all these years?"

For several minutes the two old friends became reacquainted. Finally, Pete said, "You done something illegal?

"No. I'm hiding from some bad guys who want a piece of my ass."

Leeper laughed, "I figured you was hidin' out in the old Highgrove house."

"I was. I'd like you to swing by there on the way home. See if you can tell if anybody's been there. Particularly, has anybody been out to the storage shed."

"That's all?"

"One more thing. I've got another office in the Mission Inn. Number 4F. The name on the front is Materials Analysis. Could you check it out as well?"

Pete said, "Sure, happy to do it. Anything else?"

"That will be plenty. I'm long distance, so I'll call you later at home. What time should I call?"

"Call me in an hour."

At 5:30 p.m., Greg placed the call to Pete. His wife answered and said he wasn't home yet.

He called again at 6:00 and became more anxious to hear Pete was still not home. At 6:30, Mrs. Leeper said, "He's pulling in now. Hold the line."

Pete came on with a raspy voice, "Greg?"

"Yeah, how'd it go?"

"It was the most horrible thing I ever seen. Did you know what I was gonna find?"

"What are you talking about?"

"Ralph Gould, that's what I'm talking about!" He screamed into the phone. "It's the most grisly thing - ."

"Oh God, Pete. What is it?"

"Somebody did a number on him. That's for sure." Leeper's hoarse voice sounded as if he was in shock.

"Pete, what did you find?"

"Got to the Mission Inn and the door was open. Walked in and there was Ralph staring right at me. Just sitting in a chair, staring at me."

"Go on."

"Dead as hell! Arms and legs tied to a chair, neck twisted half way around. Cigarette burns all over his face. Horrible, just horrible. Worst thing I ever seen."

The motel room started to swim in circles. Emily put her arm around Greg's shoulder to brace him. She whispered, "What is it?"

Greg took a deep breath and exhaled. "Pete, did you go to the Highgrove house?"

"Went there first. Locked up with a 'for rent' sign. Curtains pulled."

"Did you check the storage room?"

"Damn right I did. Weeds a foot high, rusty old lock in place on the door. Greg, what in hell happened to Ralph? Damn, he was ripe. Was it because of you?"

Greg paused. "I'm afraid so."

"I want no part of this, Greg. You're into something way out of my league. I'm not even gonna call the police. I don't want to be associated with this. Ralph was trying to help you. I don't want to help you and end up like him. Most terrible thing I ever seen."

"I'm sorry to the bone, Pete."

Greg set the phone back in its cradle and slumped to the bed. Emily said, "Ralph's dead, isn't he?"

"I'm afraid so. Hoyer tortured and killed him the same way he did Ken and David."

Emily hugged him tightly around the waist. "Did he find the machine?"

"No. Ralph didn't let me down. He gave Hoyer New Orleans, but neither the hotel where we were or our machine. Just when I think it can't get any worse, it does. I shouldn't have brought Ralph into this. My God, I killed him."

"Hoyer killed him. Neither of us thought Ralph was in danger."

"They're going to find us, Emily."

Emily squeezed both of her hands. "Not if we stay under cover for four months. That's all the time their Congress has allotted them to chase us down."

The Greyhound bus from Blythe arrived in Riverside a little after 1 o'clock Wednesday afternoon. On Tuesday, Emily had gone to the only beauty parlor in Blythe and had her hair shorn and dyed into an auburn pageboy. In the afternoon, she wrapped the towels around her waist that Greg had purchased. He secured them with plumber's tape and handed her the maternity clothes he bought while she was at the beauty parlor. She looked into the mirror and grimaced.

"I don't even recognize you," Greg said.

On the ride across the desert, Greg explained, "We're going to drop in and out of Riverside this afternoon. Hoyer will want me to find out about Ralph, so there'll be another communication from him in the newspaper. I want to read it. It might give us a clue about where his henchmen are looking."

"He'll find out you read it and start another misery."

"You're right." Greg sighed.

Emily said, "You're not going to read it. I am: a dark- haired pregnant woman. You can drop me off at the library as soon as we get there."

Greg nodded. "Meet me back at the depot in time to catch a 4 o'-clock bus out to San Diego."

"Why San Diego?"

"Because they're finding us too easily when we stay in hotels. I want to buy a camper and stay on the road and in campgrounds until they ease off. That means I'll need new identification, and the only place I might get it is in Tijuana, Mexico."

"You thought we'd be safe indefinitely in New Orleans and they found us almost immediately," Emily said.

"It'll be different this time. Nobody, including us, will know where we're going to be on any given day. Plus, I'm going to change my appearance tonight."

Emily was breathless when she scurried into the depot.

"You were right, the message was there in last week's paper. Here, I wrote it down. It doesn't make any sense to me."

Greg, heard about Ralph yet? Sorry about your new disfigurement, but it's going to make it a lot easier for me to find you. It was a little hard on your pregnant mother though. Why don't you be kind to her and your friends and give it up. Now that you've read this, there will just be more miseries for you, tough guy.

Herman

They puzzled over what Hoyer was referring to as a disfigurement. Greg was concerned, and stripped in one of the men's room's water closet stalls and thoroughly inspected his body.

Once on the bus, he said, "I don't have any disfigurement. He must have interfered somehow during my mother's pregnancy. Maybe he thought he'd crippled me before birth or something."

After their arrival in San Diego, they took a taxi to San Ysidro, next to the border, and checked into a motel. Greg got a crewcut in a mall across the street and purchased bottles of peroxide and ammonia.

"Don't you look cute in a crew cut." Emily complimented him. "What's the peroxide for?"

Greg put an arm around her and brushed his hand across his head. "I'm going to bleach my hair like I used to in high school." An hour later, he stepped out of a shower and looked in the bathroom mirror at his crew cut, now several shades lighter.

Emily said, "It doesn't look right with your beard."

"You're right. It's time to shave."

Emily was sitting on the bed watching the evening news on the television, when Greg walked out of the bathroom with a towel wrapped around his waist.

"Well, I found the disfigurement."

Emily looked up and gasped. A hideous purple birthmark covered his lower left jaw.

Chapter 27

It was Wednesday afternoon when the astronauts arrived in Riverside. Webster went immediately to the library to see if Greg had shown up to read the message he'd put in the Press-Enterprise after reporting to the director. His librarian lady friend said she'd not seen the Dungeons and Dragons man a second time. He met Ellis an hour later. Ellis had been to see Janet Gould. She had neither seen nor heard from Philips, but was concerned because her husband hadn't been home for days. Webster sent Ellis to check out the bus station.

Francisco arrived a half hour later, obviously shaken.

"Cody, there's a murdered man in Materials Analysis."

"What?"

"Not just murdered, but tortured first. Been dead for some time."

Do you know who it is?"

"I looked in his wallet. It's Janet Gould's husband, Ralph. It reminded me of the Birth of Time Travel story we learned in school. Remember Philips' best friend and mentor was tortured and murdered just like that the day Philip's first time traveled?"

Webster nodded.

Ellis walked up and said, "A man fitting Philips' description caught a Greyhound to San Diego three hours ago. He was accompanied by a short, dark haired pregnant woman."

"So Emily has a disguise. Let's get to San Diego," Webster ordered.

Francisco asked, "How about Gould?"

Webster said, "Our charter is not to solve 1968 murders. Our orders are very precise: not to attempt to save lives, stop accidents, or get involved with women."

Ellis added, "My instructor said we should tiptoe as to not bend a single leaf of grass for fear of influencing the future."

Webster said, "Our director wants us to trample all over the frigging grass."

The astronauts landed at dusk on the beach near La Jolla, fifteen miles north of San Diego. After checking in San Diego, the astronauts fanned out in case Greg and Emily had taken yet another bus out; Francisco went to the airport, Ellis to the train station and Webster to the taxi line. They agreed to meet at Mercury-Four the next morning.

Webster interviewed each cabbie as he entered the line. It was a little after seven the next morning when he met the cabbie that had taken Greg and Emily to San Ysidro. He jumped into the taxi. He couldn't wait to reconnoiter with his subordinates.

Greg and Emily boarded a Mexicoach bus just minutes before Webster's cab screeched to a stop in front of their motel.

Chapter 28

TIJUANA - 1968

The Mexicoach bus pulled into its terminal at the corner of Madera Avenida and Calle 7A. Greg and Emily checked their small bags at the terminal and purchased a city map. They walked west one block to the main drag, Avenida Revolución, and approached the row of ancient Detroit-built dusty taxicabs of various vintage in front of the Palacio Jai Alai. Greg leaned into the driver's side open window of the lead cab and haltingly asked,

"*Entiende inglés?*"

"*Si, señor, get in.*"

The 1953 Plymouth pulled away from the curb. "Where do you want to go?"

"I've had my California driver's licenses revoked. I need another. I've heard I can get one in Tijuana."

The driver pulled over to the curb and turned. He had thick black hairs protruding from both ears. "What you are asking for is illegal."

"I know, but I must be able to drive and desperately need a license. I can pay the going rate."

"I know nothing of this kind of business." The driver started to turn back and engage low gear.

"I'm not a trouble maker. I just need a driver's license and am willing to pay." Greg reached in his pocket and flashed a role of 20 dollar bills.

"I am not such a person. I have a family and cannot afford any problem with the police."

Greg pulled loose two twenties and pushed them forward. "I'm not the police. Could you just tell me how to locate such a person?"

The driver reached for the bills, Greg handed just one forward. "I'll give you the second one when I have met such a person."

The driver stuffed the twenty in his khaki shirt pocket and replied, "Maybe I have heard of one who has some kind of contact. I will look for him and ask. Where will you be for the next couple of hours?"

"We have nothing to do. Maybe lunch. Could you recommend a place?"

"I will take you now to the Café de Jai Alai. Eat a long lunch in the outdoor eating area. I will find you there."

It was still late morning when he dropped them at the corner of Ortega and Avenida Revolución. They drank two Dos Equis while perusing the menu and ordered antojitos of chalupas and tortas.

"Well, we seem to have made it into and out of Riverside successfully, although I'm not going to take the chance of reading about any more of Hoyer's miseries." Greg rubbed his cheek.

"I'm getting used to the mark. It doesn't matter to me a whit."

"Thanks, Emily. It's strange, but I seem to have overnight developed a lifetime complex about it. I'm sensitive to the stares and immediate averting of people's eyes. I guess, as of now, I have had it all my life. Isn't that strange."

"If it bothers you so, just regrow the beard. I thought you were very cute in it." Emily leaned over and gently kissed him on the lips. "Tell me your idea about living in campgrounds."

"If we take any form of public transportation, Hoyer's boys are going to trace us. They're very resourceful. The same applies to hotels, motels, and restaurants. The only way I can see to avoid all of them is to buy a used self-contained camper and move every day."

"Where will you buy it?"

"It all depends now on our success getting new I.D. We can't buy anything in California without registering it with the motor vehicles department. If we we're stopped for the most minor violation, we'd be busted, and Hoyer's boys would be down on us. We need for me to have a new driver's license."

"Can these Mexicans give us what we want?"

"I guess we're going to find out."

They had finished their antojitos and were stretching their second Dos Equis when the taxi driver arrived with a younger version of himself in tow.

"This is my second cousin, Mario. He believes he might be able to help you."

Greg shook Mario's hand and bluntly asked, "What can you do Mario?"

"My $20 first, señor." The driver reminded him, stuffed it next to its mate in his khaki shirt, and left before the discussion of things he didn't want to hear began.

"Come with me," the young Mexican ordered and turned away. Greg dropped the appropriate amount on top of the bill and followed. Mario opened the rear door for them on a 1955 Chevrolet, got behind the wheel and quickly forced his way into the stream of traffic on Revolución.

Mario wound in and out of numerous poorly paved and poverty-stricken neighborhoods. Greg gave up on even the direction they were heading. Mario suddenly turned into a squalid and unpaved side street. If there had once been a street sign, it had long ago been vandalized and not replaced. He braked to a stop causing a wave of dust to envelop the rusty Chevrolet and jumped out. A wave of the hand to follow, and he disappeared into an amateurishly constructed adobe brick building covered in graffiti. Emily looked apprehensively at the windowless construction and shuddered as Greg took her by the elbow and followed Mario into the dank interior.

Mario was talking rapidly and quietly to an elderly, fat and very dark Mexican sitting behind a scarred wooden desk in the front office. After a brief discussion, the fat man pushed out of the squeaky desk chair with a wheeze and walked toward Greg.

He dismissed Mario with the casual wave of a hand. Mario scurried out the door with a worried expression.

"Tell me in your own words what you want, señor."

"A California driver's license."

"This is not an unusual request. However, when you gringos hear the price, you usually decide you don't need to drive for a year or so after all. Only those in bad trouble with the necessary dinero follow through. Which are you, señor?"

"I guess the latter."

"If you're the latter, then you probably need more than a driver's license. You need the social security card as well.
Actually, we only sell them as matched pairs, because that's the way we buy them from our compadres Norteamericanos. They are not inexpensive."

"When can I get them?"

The fat man chuckled, "Isn't it always the way with you people on the run. You get to Tijuana and think we should solve your problem the moment we meet. No, señor, it is not that easy." He folded his arms across a belly that too many burritos had passed through over too many years.

"I understand," Greg replied. "What can I do to help make it happen as soon as possible?"

"We could provide you a simple fake driver's license that wouldn't make it through your first speeding ticket. There are those in Tijuana who would be happy to take your money for this worthless piece of paper. What we supply is a legitimate license and social security card with a valid number. We pay heavily for the matching pair obtained from the recently deceased. The death certificate for them is never filed, and so the buyer becomes that person. Many people in the California bureaucracy have to be paid to accomplish this sophisticated transaction."

"I'm impressed," Greg said. "How much will this matching pair cost me?"

"Eight thousand dollars. Four thousand now to assure your good faith. I must put in up-front money with the Norteamericanos.. Four thousand more when they locate a recently deceased that fits your general age and description. We will have the additional problem of your marked face."

"Don't worry about it. I intend to grow a beard to cover the mark. Your price seems high, but I won't barter if I can have it within 24 hours."

"Oh, señor, I do not know."

"I'll give you the four now, and six more if I can have it by tomorrow night."

The fat man extended his hand. "I enjoy doing business with American executives. You know how to excite a fat old Mexican. They call me El Gordo. Do not laugh. It is a name highly respected locally. It is known I will make pain for anyone who laughs. It's all you need to know me by. Now, please step into the back and I will take the photograph."

Greg shook his pudgy hand, surprisingly small for one so rotund.

"Where are you staying, Gringo?"

"We don't have a room."

"I recommend the El Rey on Díaz Mirón. It is a Mexican businessman's hotel and not frequented by rich Americanos. I will come to you tomorrow evening with or without your identification. If it takes longer, I will only charge you the additional $4,000. I am as honest as my profession will allow."

"Thank you, El Gordo." Greg was careful not to grin.

"Do not go out of your hotel once you arrive until I come. You are obviously being hunted, and I don't want you gone before I get the rest of my dinero."

"Mario did me a favor to introduce us. Will he be paid?"

"He is one of mine. You are a considerate man to ask. Do not worry about him. Until mañana, vaya con Dios."

WEBSTER STOOD in a muddy parking area a few hundred yards across the border arch on Mexican soil. He had been interviewing every driver that pulled into the lead position of the taxi line for over an hour.

"Maybe I have, señor," a dark, high-cheek-boned driver answered.

"Your description is correct for a couple I picked up yesterday morning at the Palacio Jai Alai, except the man had blond hair and a purple birthmark on his face."

Webster was puzzled for a moment and then decided it could be a Philip's disguise. He climbed into the rear of the taxi. "Take me to where you dropped them off."

The taxi driver didn't move. "My cousin took them to their final destination."

Webster asked, "How do I find your cousin?"

"He is an expensive young man to find, señor."

Webster leaned forward and flashed a one hundred dollar bill. "This expensive?"

The driver snatched the bill and popped the clutch in one motion. "I will drop you at the Café de Jai Alai and go find Mario. I warn you, he will not want to help you. He will be very expensive."

GREG and EMILY had been content to spend the last 24 hours in their hotel with food delivered by the bellboy from nearby taquerías and loncherías. They entered the hotel's tiny cantina as the cantinero lit a hanging lantern to illuminate the bar.

Emily was seated facing the open door leading into the hotel lobby. Greg ordered bottled water from a pretty señorita. Suddenly, Emily put her hand on his arm and whispered, "Greg."

The waitress moved to the bar and Greg asked, "What is it?"

"It's Webster; I'm sure of it." She nodded toward the lobby.

"Where?"

"Talking to the registration clerk."

"Oh, God, it is!" Greg spun his head around to find another exit. There wasn't one. The clerk pointed in the direction of their room on the second floor. Greg whispered in Emily's ear, "Let's get behind the door and make a run for it when Webster goes to our room."

Greg watched Webster hand the clerk a bill and start up the stairs. The clerk yelled after him, "Señor, you might try the cantina first."

"I've got to run for it, Emily. I'll meet you at the Jai Alai café in an hour," Greg whispered and lunged through the door. Webster spun around and bounded down the stairs. Emily dove into his path, the two of them slammed together and fell into a heap onto the adobe tile floor.

Greg was on the sidewalk running flat out darting between pedestrians and street vendors. Webster pushed himself to his feet, but Emily got her arms around his ankles and hugged them with all her strength. "Let go!" Webster yelled and reached down with both hands and twisted her arms away.

By the time he was out the door, Greg was at the cross street running through traffic against the light. Webster tore after him. Emily ran out to the sidewalk and saw Webster running into the cross street. A taxi pulled up in front of her and El Gordo waddled out of the front passenger side. Two husky young bodyguards were in the back, one about to pay the driver.

El Gordo looked at the horror on Emily's face and asked, "What is wrong, señorita?"

"A man is chasing Greg that way!" She pointed at the corner.

El Gordo dove back into the taxi and barked a Spanish order. The taxi spun in a u-turn in front of the on-coming traffic and bounced over the curb. "Vámonos!" El Gordo yelled into the driver's ear and pointed toward Webster's running figure entering an alley marketplace. The taxi skidded to a halt just as Webster caught up with Greg and dove at him in a football tackle. Greg slammed into a street vendor's display and leather wallets and purses flew in all directions. El Gordo bellowed orders to the two young men. They sprang out of the car and the larger of the two drop-kicked Webster in the head with a sharp-pointed rattlesnake cowboy boot. Webster collapsed.

El Gordo spat out additional orders and then reached down and helped Greg to his feet. "Can you walk back to the hotel, señor?"

Greg dusted himself off as the outraged street vendor vented his anger at El Gordo. El Gordo shoved a wad of Mexican pesos into the vendor's hand, took Greg by the shoulder and pushed the two of them through the crowd that had assembled.

Greg saw the bodyguards lift Webster into the taxi and drive away. Greg and El Gordo retraced Greg's steps to the street and found Emily looking into the milling crowd.

"Thank God you're okay," she sighed. "Where's Webster?" El Gordo answered, "The aggressive young man? He will be detained until you are safely out of my country."

Back in the hotel lobby, El Gordo's eyes took on a menacing gaze. "I am ready now to complete our transaction."

"Come to our room," Greg said.

Once inside the room, El Gordo produced a legitimate driver's license and social security card in the name of Edward Best. El Gordo explained they had been in luck, although Mr. Best had not. He had been drinking prior to departing Indio for El Centro, had fallen asleep at the wheel and rolled over several times into a dry desert gulch. The highway patrolman, first at the scene, recognized Mr. Best fit the general description provided by El Gordo. The patrolman removed Best's driver's license and social security card before calling the ambulance. He told the ambulance driver that Mr. Best had been alive upon his arrival, but knew he was dying and requested to be taken to The Resting Place Mortuary in El Centro. The Resting Place was a member of the conspiracy ring.

Greg thanked El Gordo and handed over the balance of $6,000 in one hundred dollar bills.

"Now get out of Mexico pronto, gringo. We will hold this aggressive young pursuer of yours for tonight. But it's too dangerous for us to hold him past morning. He will be drugged and dropped off at the border. You have a head start of one night. Good luck, gringo."

Chapter 29

After Tijuana, Greg and Emily had taken a bus to El Centro hoping to convince Webster if he picked up their trail they were going to reenter Mexico through Mexicali. Instead, Greg used his new Edward Best identity to purchase a two-year-old white Ford pickup complete with a camper shell. The camper included an over-the-cab double bed, small kitchen and dining area, and a toilet.

Emily purchased a Double Eagle Guide to Western State Parks. Greg stayed within the confines of the camper whenever they were not on the road. His marked face was too memorable, and he feared Webster's resourcefulness in tracking them.

The first day they back-tracked across Route 78 to the little hamlet of Julian and stayed overnight at Agua Caliente County Park campgrounds, located on forested slopes in the Peninsular Mountain Range. Emily left Greg in the camper with his slide rule and calculations and had her hair re-dyed back to honey blond in a local beauty parlor. After dark, Greg soaked his sore back in the local hot spring.

In the days that followed, they crisscrossed the state, staying in out-of-the-way campgrounds: Red Rock Canyon off State Highway 14 east of Bakersfield, the Forest Service Campground near Mammoth Lake on Route 203, and Ahjumawi Lava Springs northeast of Redding. It was a pleasant lifestyle. Emily would buy the groceries and they would cook together, take evening walks under the stars, get up at dawn and fish when they were near water. They avoided other campers and paid cash for everything. They never used the same name twice when checking into a camp. After a few weeks, Greg's beard once again covered his face and a haircut removed the grown-out dyed hair.

After two months, Greg gave up his quest to extract the secret of future travel through calculations. There were intricacies he could not decipher on a slide rule.

Two days before Christmas, they pulled into a state campground south of Pismo Beach, California. Emily had read about the local clam digging and was anxious to try it. Greg sauntered into the office and waited patiently behind an elderly retired couple intent upon telling the ranger the full extent of their travels. Greg wandered across the room to a bulletin board and was thunderstruck!

He backed toward the door as the ranger looked over the shoulders of the couple and said, "I'll be with you in a minute."
He started to turn back to the couple and then seemed to do a double take. Greg hurried to the pickup.

"What's wrong? You look like you've seen a ghost."

"Worse, I've just seen us. There was a reward poster in there, that has photographs and descriptions of us."

"Ohmigod! What did it say?"

"There's a $20,000 reward for anyone who finds us and a manned, 24-hours-a-day phone line."

Greg was backing out of the parking space, as the ranger stepped out the front door and stared at them.

Emily asked, "Did he recognize you?"

"It looks like it." Greg skidded half way out into the Dolliver Avenue intersection and careened toward downtown Pismo Beach.

"Slow down, honey, we'll be in a bigger mess if you're pulled over."

Greg took a deep breath and decelerated. "We have to get off the road."

They drove through the little town of Pismo Beach and turned north on the frontage road passing restaurants and motels.

Within three miles, the frontage road ended at a cross road leading to the Pacific Ocean a few miles west. A small road-sign pointed the way to the San Luis Pier. Greg had an inspiration as they were passing Avila Hot Springs and jammed the brakes to the floor and backed up into the Avila parking area. His voice was hoarse with anxiety.

"Honey, this camper is a dead giveaway. We have to put distance between it and us. I'm going to leave you here while I ditch it."

He pulled in front of the old mineral hot springs office and read the fading particulars on a weathered wooden sign to the right of the office door. Private campground, welcoming overnighters or by the week, individual private tubs by the hour. "Sign in and tell them your husband will be joining you."

Greg drove west to the ocean a few miles away. San Luis Bay came into view. The old pier was highlighted by the evening sun sliding behind storm clouds. Greg drove past rows of docked fishing and pleasure boats. He pulled to the left of the pier entrance, sat quietly in the darkened cab and looked down the length of the pier, now closed and locked for the night. A single, weathered, open building was at the apex, and a small, fishing boat ice house with moss-encrusted fishing nets strung along one side.

As his eyes accustomed to the rosy-gray twilight, it started to sprinkle. Only one private party seemed to be underway inside a small yacht anchored yards away. Greg got out of the cab and walked to the pier gate. It had a weighted blockade arm secured with a rusted lock to a support pier post. He looked down upon the luminous activity of plankton in the bay water. He walked to the Ford and took a large common screwdriver from the glove box. He returned and leveraged the screwdriver behind the lock's hinge and pried. It tore loose from the wood and the gate swung slowly away. Greg returned to the camper and climbed in.

When the pier was completely engulfed in the mountain's shadow, he took one last look around. He heard only the light rain except for something spirited and barely audible oozing from a radio on the yacht. He rolled down the driver's side window and unlatched the door. His heart was beating wildly as he started the Ford. He backed up with the lights off and then drove forward slowly onto the rickety pier. Two thirds of the way out he suddenly accelerated and jerked the wheel to the right. The Ford crashed through the flimsy railing and momentarily poised in mid-air above the inky sanctum below. It hit the water like a playful great whale and immediately started to sink.

Greg shoved his shoulder aggressively against the door and rolled out into the icy black water. He kicked away from the sinking camper and surfaced beneath the pier, coughing up the briny sea water. After getting his bearings, he swam clumsily under the pier back toward the shore, testing the depth with his foot until touching the bottom a few yards from shore. He stood quietly and waited anxiously to see if the commotion had attracted the attention of the yacht party. Satisfied that no one was coming, he climbed out and scrambled up the rocks to the parking area above. The bay swallowed the camper without a trace. Greg started the three-mile walk, now in a steady rain, back to Avila.

Chapter 30

Webster awakened with a splitting headache, sitting against a brown stucco security wall surrounding a Tijuana residence. He asked directions to the border from a passerby and found he was only a few blocks away. The U.S. border guards looked at his filthy appearance in disgust as he walked through the Customs declaration line and then navigated the long, sloping footbridge over the incoming highway into Mexico.

He caught a local mini-bus from San Ysidro to La Jolla. Francisco and Ellis were waiting. They rented a car and returned to San Ysidro to pick up Philip's trail as Webster slept fitfully in the back seat. They found the scent to El Centro within an hour and pulled into the dusty little farming community by dusk. The trail promptly ended. The Mexico border was just a few miles away and they proceeded to check out the northern Mexican States in vain for several weeks.

At a loss, they returned to El Centro and all three spread out to interview every farmer, barber, clerk, waitress and policeman. Ellis found a used car salesman who remembered Emily and a guy who stayed in the background and signed the papers on a white camper. No, it wasn't he who had sold it to them. At this commission-only lot, salesmen came and went with regularity.

Within a day, a campground manager from the Agua Caliente Campground near Julian remembered a pretty brunette in a white, nondescript camper staying over one night several weeks earlier. He remembered her because it seemed odd for such a pretty lady to be traveling alone; also because she was a blond when she pulled out. No, he didn't have a record of over-nighter's license plates.

Webster had several hundred wanted posters printed and mailed them to campgrounds the length of the state, promising each ranger $100 just for the posting. Three days later his phone started ringing off its hook. Ellis was canvassing the state to the north and Francisco to the south. Several obviously false reports intermingled with probable valid ones.

His eleventh caller was a serious sounding ranger from Pismo Beach, one hundred miles north of Santa Barbara. He was positive the couple he'd seen were the wanted ones. He described the bearded man's reaction to seeing the poster. Webster had to wait an hour until Ellis and Francisco made their routine call-ins. He told them to drop everything and get to Pismo. He would move Mercury-Four and command headquarters to the beach at Pismo that night.

Chapter 31

G reg shook off the rain on the Avila porch and then bypassed the front desk and approached the bath house attendant, a roan-haired woman with wrinkled alabaster skin. She unlocked the bath hallway and escorted him to Mrs. Webster.

Emily was sitting waist deep in the iron cast tub for two, her breasts and shoulders exposed. "Am I ever happy to see you! Did you get rid of the camper?"

"It's at the bottom of the bay."

"You're soaked."

"It's raining."

Greg quickly undressed down to his shorts and slid into the water. Emily looked at his shorts ballooning in the bubbles.

"Are those air bubbles, or are you just happy to see me?"

Greg smiled weakly. "Bubbles, I'm afraid."

The 30 minutes in the hot sulfurous water revived him. He said, "Ready to go?"

"I have been for an hour. What are we going to do?" Her hands were tightly folded.

Greg was quiet while he dressed and then said. "We'll have to buy a ride from one of the campers outside. Let's go."

He opened the door for Emily who stopped in mid-stride and gasped. The roan-haired female attendant had been replaced behind the counter by a tall, handsome young man.

"Good afternoon, Dr. Philips," said a husky voice whose timbre resonated. He raised a weapon above the counter.

"Cody Webster," Greg sighed.

"I'm afraid so." Webster dialed out. "Bill Ellis' room please." He waited. "Okay, I'd like to leave a message. Have him call Cody Webster the minute he gets back. Here's the number."

He replaced the receiver. "You're a tough man to find, doctor." His eyes hadn't wavered off of Greg throughout the call. "We're going to wait here until I get help. My buddy is an hour away at a hot springs motel in Santa Maria."

"And then?"

"He'll hold you while I get the director. He wants to bring you in himself."

Greg's mind was racing. "How did you find me?"

Webster studied the venerated scientist and admired his relaxed demeanor. His own arm pits were soaked. Philips was more august in person than his photographs showed in the history books.

"We received several reports from campgrounds you'd stayed in. Hot springs were the common denominator." Webster was touched to see Emily wrap her arms protectively around Philips.

"Before entering any particular time, we're required to read every available piece of computerized information. The agency computer does a search for known diseases during the time and we're immunized accordingly. I was afraid we were going to look in 1968 and forward for a long time."

"You know that isn't true," Greg said. "You can always take one of your shortcuts."

Webster raised the weapon above the counter and rested his elbow before asking, "What are you talking about?"

"Your misery policy."

"I don't know what you're driving at."

"Torture and murder."

"Torture and murder? Us? You're the murderer."

"Me?"

"The trail of bodies you're leaving behind. Your collaborators in

1983 and your high school friend in 1968. If there wasn't such adulation for you in 2030, our director would have eliminated you. As it is, he's received congressional approval for us to stop you any way short of killing you."

Greg rolled his eyes. "Do you seriously believe I killed my friends?"

"Who else? It was commonly believed your co-workers in 1983 had been murdered by druggies after your departure into time. We all read about it in our history classes. It wasn't until Ralph Gould's recent murder here in the same manner that the 1983 murders were reevaluated."

"Who did the re-evaluation?"

"The director."

Greg chose his words carefully and slowly, "Hoyer's the serial killer, and he's the one accusing me. Incredible! What motive did he say I had?"

Webster squirmed. "For Dr. Hoard and his son, it was to stop them from getting any of the credit for your work on black holes and time travel. And Mrs. Gould provided me with your motive for her husband's murder, even before you'd committed it."

"What are you talking about?"

Webster shifted his weight. "I tracked the Goulds to Hawaii and talked to Mrs. Gould. She told me how she'd dumped you in 1953, and you never got over it. You showed up again recently and were harassing her. She wanted to tell her husband but was afraid of the outcome. Obviously she was right."

"That's stupid!" Emily shook her head. "It was just the opposite."

"Her husband's murderer has to have been Dr. Philips." Webster shifted his weight again. "He's the only person who was in the same place and time for all the murders."

Greg said, "I wasn't the only person. Hoyer was in both places, as

175

well. He's your murderer." Greg described the events of his 1953 trip in time and finished with, "I found Ken and David's son's bodies when I returned to 1983. It was then I realized how Hoyer had found me so rapidly. He'd told me it had taken years to locate my predecessors, Honjo and Pratt."

Webster's gun hand slumped and beads of sweat popped onto his forehead. "You knew about them?"

Greg nodded. "Both victims of your director. Honjo crippled and Pratt murdered."

Webster looked perplexed. "I can't believe this."

"Emily and I were in the Weston hotel in New Orleans when Hoyer tortured and murdered Ralph for our whereabouts. Call the Weston."

Webster picked up the telephone receiver and called Louisiana information. He put his hand up palm forward toward Emily, who was talking excitedly to Greg as he asked for the number of the Weston Hotel in New Orleans. Minutes later, he replaced the receiver. "Your story checked out. I wouldn't have called, except I've wondered how the director knew you were in New Orleans."

Greg said, "Then you know I couldn't have killed Ralph."

"I don't know anything except you've thrown some questions at me. How about Hoard and his son?"

Greg had an inspiration. "After I found Ken and David, I left a note addressed to Jeff Southwell of the CIA telling him Hoyer was the murderer. I left it out for Hoyer to find."

Webster commented, "Hoyer told me about it. You were going into ancient history."

"So he did get the original," Greg mused. "I'd made a copy, however, and hid it in the bathroom storage cabinet. The last call I made before coming here was to Southwell to tell him to pick it up the next morning. All you have to do to verify this is travel to 1983 and talk to Southwell."

"I couldn't do that if I wanted. All of our trips into time are authorized by Congress and under computer control," he paused, "except for the director. He's the only person on Earth who can get presidential approval for an unauthorized trip."

"So you're only authorized to travel to and from 1968 to 2030?"

"That's right."

"Then go to 2030! Southwell's probably still alive."

"I can't do that. I have to report all of my activities to the director." Webster set the gun down on the counter and spread his arms.

Greg asked, "Where's your saucer?"

"I can call it up whenever I want."

"You've got an hour until your friend arrives. You could make a trip to 2030 just long enough to call Jeff and confirm what I told you."

Webster said, "You'd be off and running the instant I was gone."

"Take me with you."

Webster inhaled deeply, casting off tension. "I don't know."

"Damnit, Webster, you're talking about my life."

"Dr. Philips, have you always had that facial birthmark?"

Greg pushed his beard up to expose his skin. "You mean this? Hell, no! It suddenly appeared after we left New Orleans."

Webster slumped slightly and mumbled, "That must be the disfigurement the director was talking about."

"When?"

"In the personal column he had me place in the Riverside paper." Webster brought his hand to the corner of his mouth as if debating to peel indecision from his lips. "Miss Johnson, you'll have to stay here. I'll set the timing of the trip so we'll be back in 30 minutes. Come on Dr. Philips before I change my mind." Webster pulled the now familiar Mercury-Four activator out of his pocket and the gleaming saucer materialized on the grassy parking lot. The hydraulic-hinged door raised and Greg entered first. He noted three seating positions like three spokes of a wheel all facing the hub. They buckled in and Greg watched Webster punch data onto the screen. All three of the first monthly green squares along the top rectangle were pulsating in dark green.

Moments later, Greg felt the gravitational pull toward the center of the craft. He braced himself as everything surrounding him stretched out into a colorful kaleidoscopic maelstrom toward the hub. He regained consciousness a moment later and found the craft in a hovering position. Webster was reaching for a joy stick while observing a three dimensional image on the screen of a fenced field directly beneath them. Webster maneuvered the craft away from the fence and lowered it onto the field and powered down. "I shouldn't be gone more than a few minutes."

As soon as the hydraulic door closed behind Webster, Greg unbuckled and dropped to the floor. With a coin he unfastened the Zeus connectors holding aluminum skins in place over the center column hub. He studied the exposed optics and noted the position of the electron and ion beam optics. He had barely replaced the skins when the door swung down and Webster reappeared.

"I talked to Mrs. Southwell. Her husband was killed in a clandestine CIA operation 50 years ago."

"Add another victim to Hoyer's list."

Webster said, "We don't know that. This trip didn't prove a damn thing. We're going back to 1968."

"Just a minute, Webster. If Hoyer murdered Jeff when he arrived at Ken's lab, it means Jeff never found the stuff I left him."

Webster asked, "What stuff?"

"The note I told you about, and a 1953 newspaper clipping from me to Ken that had arrived that morning. Oh, and of course, a roll of 35 mm film. David had taken several pictures of my departure through the time machine window."

Webster said, "If he was taking pictures of your departure, you sure as hell couldn't have killed him. Where can I find the film?"

Greg grinned. "I buried it in a large aspirin bottle in Ken's bathroom storage cabinet."

"Let's go take a look."

"The old building would be long gone by now."

"On the contrary. It's a national shrine with daily tours."

"I'm impressed."

Twenty-five minutes later, Webster returned in a shaken state. "Let's get the hell outta here!"

"Was the aspirin bottle still there?"

"Yeah." Webster fired up the Mercury-Four and punched data onto the screen. Greg gripped his seat's arms as the horizontal gravity began its irresistible pull. In the next instant, he asked, "Where are we?"

"Back at Avila in 1968. I'm letting you go."

The door raised and they climbed out. Webster pointed the activator at Mercury-Four and it vanished. Emily was standing in front of the office door. She ran to Greg and threw her arms around him.

"Let's get inside," Webster ordered.

"Did you talk to Southwell?" Emily anxiously asked.

"With his wife. Dr. Philips can tell you about our trip later. I know now my director is the murderer and intends to kill Dr. Philips as soon as he's alone with him."

"What did you find?" Greg asked.

"I showed the ranger my credentials and was given a free run of the place. It isn't every day they have a time astronaut take the tour. I opened the cabinet and found an old aspirin container. Inside was your film, note and clipping."

"You've got them?"

"You bet, I just refilled the aspirin bottle and sauntered out. They're going to be my insurance if the director finds out I helped you escape."

Emily asked, "Can't you blow the whistle on him?"

"All I've got is evidence proving Dr. Philips isn't the killer. It's too circumstantial to nail Hoyer."

"How did a person like him climb so high?"

"Hoyer's a workaholic who loves his job. No family or outside interests I know of outside of boxing in the NTSA gym. The president knows he can count on him to get a job done."

"So you can't go up against him?" Emily pondered.

"He'd discredit me immediately. No, all I can do is … ."
Webster was interrupted by the ringing of the telephone.

"Hi, Bill, thanks for calling back. No, nothing new. How about at your end?" He chatted for several minutes and told Ellis the hot spring theory was a failure, and he was going to have the three of them reconnoiter and look elsewhere. "Give Tony a call. We'll meet this evening in Santa Maria."

He replaced the receiver and turned to Greg and Emily. "You better get out of here. Which direction will you head?"

"Back to Southern California."

"Then I'll point my guys in another direction, but I won't be able to keep them off your trail for long. Hoyer's never going to give up on you. It's like a vendetta."

"Did he tell you Greg whipped him in a fight the first time they met in 1953?" Emily asked.

"You whipped Hoyer? Nobody will go into the ring with him. He's savage. You whipped him? Amazing!"

"Probably just lucky." Greg smiled and reached out for both of Webster's hands. "We owe you, Cody."

"Where's your van?"

"In the Pacific ocean."

"Take my rental car. You have almost no chance."

Greg smiled. "I'll dance on your director's grave."

Chapter 32

It was after midnight when Greg and Emily pulled into the driveway of their Highgrove house. Greg found the front door cracked
open, the wood torn around the lock. He flicked the living room
light switch on. Nothing happened. He realized power had been turned
off, probably for months for non-payment. Exhausted, they groped
their way to the bedroom and slept. At morning light, Greg checked
out the backyard and concluded the astronauts had made it to the house
but not water tower.

He saw that the electric power meter was stopped and locked with
a wire and ceramic seal. He cut the seal with dikes and activated the
power.

Once in the tower, he saw his machine sitting eerily in the center
of the room. He gathered his tools around him and started disassembling the ion gun. He said to Emily, "I took a close look at the optics'
configuration of Cody's Mercury-Four. I know where I went wrong."

Greg continued through the day, stopping just long enough for
lunch. Emily delicately commented, "Your machine doesn't look like
the Mercury-Four, dear."

Greg looked up from his paper plate and stopped in mid-bite. He
could see the strain was taking its toll on Emily and he framed his response carefully. "My prototype doesn't have all the bells and whistles
the Mercury-Four has, honey, but I'm confident it'll do the job."

To convey his confidence, he lazily stretched until his joints
cracked with a leisurely twist of his body. He worked through the night
until the cold morning light tentatively fingered its way over the surrounding orange grove. He'd disconnected the external input power

them and Emily squeezed her hands. Each of them was now buried in their own supplication to God.

In silence, Greg set the controls to settings he felt would take the machine and its contents into the future a matter of minutes to a position a few yards into the back yard. He carefully noted the time.

They sat together as darkness arrived. A feeling of rancor overcame Greg. He had failed. There was nothing left to try. They went to bed with the feeling of hopelessness they'd had on their last night together in 1953. They spent a sleepless night wrapped in each others arms.

In the morning, Greg looked out the back door to an empty back yard. "We may as well hit the road. There apparently is more to the Mercury-Four than I realized."

"Let's give it the morning, honey." Emily poured them cups of coffee. "This is our only chance."

Her pragmatism impressed Greg, and he hugged her. "We made a good run for it, honey."

"You bet we did." She kissed him heartily on the mouth. "And if Hoyer comes and tries to hurt you, I'll kill him myself!"

Greg grinned, "You are one helluva woman, Miss Johnson!" He lifted her off the floor and swung her around the kitchen.

"Greg, look there!" He sat her down.

"In the yard. The machine just materialized in the yard."

Greg noted the exact time on his watch.

They ran outside and looked in awe for some minutes. Then, they returned to the kitchen and made a list of things they were going to need: Emily was to take the balance of her stock certificates out of Citizens Bank, cash a check and request new 1968 one hundred dollar bills. Then, she was to visit Mission Jewelers and spend whatever old money she had left on jewelry. While she was taking care of their financial needs, Greg would prepare the machine for the trip.

·

Emily returned to the house after five that afternoon, her mission accomplished. By the time Emily arrived, Greg had moved the control panel into the machine. He spent the afternoon carefully measuring the distance the craft had traveled down to the fraction of an inch. He also calculated the exact time to the second that had elapsed from the moment he set controls the previous evening until the craft rematerialized.

The wave of fear in Emily's eyes subsided as Greg explained his plan. He had extrapolated the elapsed time of last evening's time travel into a trip to the year, 2035. He had selected that year for two reasons: they would be five years beyond the time the astronauts had given up in their search for Greg, and a safe arrival would mean there had been no time splits caused by their capture in the five years that would have elapsed in the astronauts' relative time. Secondly, he had not forgotten his vow to Ken and David Hoard.

The arrival location was more of a problem. The only area of the United States Greg was familiar with that might not be populated in 67 years, would be the Colorado Desert. Greg calculated the parameters to set them down in the vicinity of the tiny hamlet of Desert Center, 50 miles west of Blythe, sometime in December. He knew they would not survive in the blistering sun of a summer day. He washed out an empty bottle and filled it with water. Within minutes, they stood at the entrance of the machine.

Emily asked, "Where do I check my luggage?"

"No luggage allowed."

"Maybe I should have taken United."

It was an unreal time for them. Emily stared ahead in her seat - transfixed. She was held in the grip of an imperative larger than herself. Greg shared her wave of fear. He changed his thoughts to Hoyer. He quietly promised, "I'm coming for you, Hoyer."

With steadiness of hand, he started the vacuum sequence and then set the optics parameters. Every muscle in his body tightened.

Chapter 33

G reg's brain reeled as he returned to his senses. He flung off the restraining belt and looked across at Emily. Her eyes blazed into his. Moments later, they were outside the craft in night air that was crisp, cool, and clean. They stumbled off with the bottle of water into the vast desert.

It was very dark. The outline of a mountain range enveloped in the darkness loomed before them. Emily said, "My God, we've done it." They hiked for half an hour into the Chocolate Mountain range. From an elevated vantage point, they scanned the night horizon and spotted a dim light in the distance.

They walked toward the light and watched the magnificence of sunrise that only a desert can offer. The sun made a burst onto the horizon and quelled their shivering. They came across a gravel road and a few minutes later, a truck rattled toward them. Greg posed in the classical hitch-hiking position.

"Climb in," The driver said. "What are you doing way the hell out here?" He was a black man with an inquisitive smile.

Greg paused. "We were prospecting for geodes and our vehicle broke down. Where are you heading?"

"Into Blythe. Carry'n gypsum from the Midland mine. It's 'bout 20 miles to Blythe. You can rustle up a tow truck there."

"This truck has the quietest engine I've ever heard." Greg commented.

The driver turned his head a little in Greg's direction. "Engine? What you been riding in? The old six and eight cylinder gas jobs?"

"Yeah," Greg shifted uncomfortably, "it's been a while since I was in a truck."

"Well, this is powered with a hydrogen rotary engine."

"When did they first go to the hydrogen rotary engine?"

"Where you been man? They've been around 20 years, since 2015!"

Greg settled back in the seat smiling and nudged Emily. She grinned and returned the nudge. As the truck descended off the mesa into the Palo Verde Valley and the outskirts of Blythe, they started passing a mixture of old familiar farm houses and new startling different ones that were pyramid shaped, blanketed in what Greg assumed were solar photovoltaic cells and copper head solar collectors that looked like sky lights. Each pyramid house looked identical and included an external water wheel.

Once in Blythe, the truck driver pulled over to the curb across from the Greyhound Bus depot. They thanked him and got out.

Greg approached the ticket counter while Emily looked over the magazine rack. "How much is it to Riverside for two?"

"$220."

Greg looked in awe at the inflated amount, peeled off three of the crisp 1968 one hundred bills, and pushed them across the counter.

"What's this for?" the Greyhound ticket vendor, a totally bald and petite man in his early 40s, asked.

"Our tickets."

"Your tickets? I can't take this. Where ya been, man? The use of cash has been outmoded for years. Where's your personal money card?"

Greg pulled back the bills. "I'm sorry. I guess I'm out of it. I've been in a sanitarium. What exactly is a personal money card?"

"You've been out of it quite awhile, mister. I'm surprised the sanitarium didn't prepare you better for the real world. Every citizen has had his own money card since the establishment of the international computer banking system in 2020. It's a card that is used by every citizen, in all but the developing countries. It's a tool that includes your driver's license, critical medical information and X-rays, passport.... It allows automatic payment of household bills or even loan applications. Your money is totally obsolete."

Greg asked, "How do I go about getting the card?"

"If you're a California resident you'd go to the Riverside County Courthouse and register. They'll prepare you a card on the spot."

"You mean I can't even buy a meal until I get one?"

"I'm afraid not. You'd best get to the courthouse in Riverside."

Greg said, "How the hell can I do that if I can't buy a bus ticket?"

The ticket vender squirmed. "I've got an idea. The night shift guy is a collector, you know, like old stamps and money. That money you showed me looked in remarkably good shape. It may be worth the price of a couple of tickets to Riverside for him. He'll be on at six. Otherwise, I don't know what to tell you."

They wandered around town and by evening were ravenous and very concerned. This brave new world posed an immediate problem for them: hunger. Greg watched the little ticket vendor explain his dilemma to a younger man of Philippine descent coming on duty. He walked across the lobby to the booth.

"Tom explained your problem. Sorry you're so screwed up. What year money you got?"

Greg shoved the roll forward.

The vendor spread out the bills and scrutinized them carefully. "Boy, I've never come across brand new money like this. It's over 60 years old and never been circulated. Pretty amazing."

"Can I use some of it to get to Riverside?"

"I guess I can buy you a couple of tickets with my PMC. Yeah, I guess it's worth it to me." He carefully collated the bills and placed them in an envelope.

"You mean you want them all for the price of two tickets?"

"What the hell man, it's nothing more than Monopoly money. It only has value to a collector like me. If you want it back, here." He shoved the envelope forward.

"No, it's a deal. Just get us to Riverside."

The bus pulled out to the west of town and followed signs to the freeway. Greg and Emily had selected seats immediately behind the driver. At the freeway entrance, the bus slowed as the driver punched data into a computer keyboard and then removed his hands from the wheel and sat back.

A recording came on for the benefit of the passengers:

"Welcome aboard the Greyhound Computer Express Van transporting you from Blythe to Riverside, California. Your van is under interstate global positioning satellite control. Portal to portal time will be one hour and twenty minutes. Your van has been assigned to lane number three. Your speed will be 131 miles per hour. Seating capacity is fifteen. The van is manufactured by General Motors and the power plant is a three-speed electric motor from General Electric. The motor is powered by ten, 20-cell nickel-iron internal mounted batteries and 90 roof mounted nickel-cadmium solar batteries. We hope you enjoy your trip and thank you for traveling Greyhound."

The recording ended and elevator music began. Greg turned and looked at the traffic occupying the lanes on either side. They were electric cars, in which the driver of the one to their right was already asleep in a reclined position. The driver to their immediate left was staring at a screen with an idiotic grin on his face. Greg leaned forward and nudged the bus driver who was about to drift off.

"Can you tell me what that driver beside us is looking at? He seems to be really enjoying himself."

The driver rotated in his seat to face Greg. "You've never seen 'virtual reality?' "

"I've just got out of a sanitarium. I've been there most of my adult life. Virtual reality? Like being a participant in the scene you're looking at?"

"You got it. Turn on your screen on the back of the seat in front of you." He turned back around. Greg tapped him again. "One more question? What's the date?"

"Saturday, November 20th, 2035." He slumped back into a dozing position.

A virtual reality screen suddenly surrounded them with an automatic, stereoscopic liquid crystal display. The screen was covered full length with both lenticular and biconvex lenses.

They sat together transfixed for the balance of the trip into Riverside watching a human interest piece showing the lunar life of a dozen settlers. It included a tour of the airlock, thermal radiators, solar arrays, life-support systems, communications gear and the transport vehicle used to rotate settlers on 45-day intervals from the moon space craft to the campsite.

Greg nodded to the driver as he helped Emily off the bus. They walked into the downtown area and found the Main Street was gone, replaced by a decrepit and obscene mall inhabited with sleazy bars, live sex shops and innumerable pawn brokers. They spent the night huddled in a Park. All the benches were taken by hordes of homeless people. They talked with the other indigents about their plights. The common denominator was a lack of, or ability to obtain, the coveted PMC. Survival was shortlived, unless you begged or were a thief. A local pawn shop broker was infamous for trading stolen goods for limited use of his PMC card.

They were the first to enter the courthouse as it opened its doors Monday morning. They apprehensively approached an officious woman who listened disinterestedly to Greg's now much embellished sanitarium story.

"Give me your name and birth date. Put your hands on the counter and I'll take you finger print computer profile."

She slid a form across the counter. "While your fingerprints are being processed, fill out your assets to be applied to the card account."

Greg put his hands in his pockets in a flood of uneasiness. "I don't have any assets."

"Thought you wouldn't." The woman's nostrils flared at Greg's unkempt appearance. "You can't get a card unless you have something to declare."

In a gesture of bravado, he returned her superior attitude by thumbing his nose. By mid-afternoon they located the pawn broker. Their anxiety had been heightened when they realized there was no survival in 2035 without a PMC card. Their only survival hope was to find a sympathetic and trusting friend of Greg's still alive. That meant traveling to the San Francisco peninsula.

Greg spread some of the jewelry Emily had purchased in 1968 on the counter. He negotiated weakly with the broker for two train tickets north, a hot meal before leaving, and a place to clean up. They embarked with a regained feeling of confidence.

Upon boarding the train, a passenger recording explained the train was supported by frictionless, levitational bearings and was propelled by on-board magnets powered by passing through battery-powered coils along track guideways. A computer in the engineer's compartment communicated with controls along the train's pathway to activate the coils sequentially to provide a traveling magnetic field in a domino effect as the train passed by them. Except for short hops to sparsely populated areas, airplanes were in disfavor.

The train pulled into San Jose at 6 o'clock Monday evening and they collected a transfer to the electrical muni-service to Palo Alto. Upon arrival at the University Street station, Greg entered a phone booth and learned how to operate the information computer. He entered in four names in succession without luck. He found the name of a friend on the fifth try.

Greg exited the phone booth in high spirits, forever the optimist. "None of the friends I worked with at SRI are listed, but Donna Buckley is! She still lives on Bryant just a few blocks from here."

"You're going to scare the bejesus out of her!" Emily commented as they walked east on University.

Chapter 34

e've reached a dead-end, sir." The three astronauts were seated around
W the director's desk. Webster continued, "After Pismo
Beach, we lost them. We revisited their house in the little
community of Highgrove and found evidence they'd been
there awhile and then vanished."

Francisco interjected, "Like they'd gone off in time again."

Hoyer asked, "Had they?"

"We checked out that possibility," Francisco answered, "Philips
had a 100 KV electron microscope in his downtown office where he
murdered Gould."

"You're on the right track, Tony. Philips had two EMs, the low
voltage one to throw us off the track and a high voltage one someplace
else."

Webster said, "I'm not so sure, Philips made a lot of money in the
market. More than enough to be murdered for. I wouldn't
be surprised that he's buried in one of the orange groves."

Hoyer said, "Ready to throw in the towel, eh, Webster?"

Francisco said, "We could return to '68 and check out the EM companies for high voltage EM sales but our four months are up."

"That's what you and Bill are going to do, Tony. I'll get authorization for more travel. Webster will stay behind with me, since he's
so anxious to sweep Philips under the rug. Try to hit as close as possible to the day you think Philips must have arrived back in Riverside."

2035

GREG PUSHED THE DOOR BELL a second time and heard footsteps approaching. The door cracked open and an elderly white-haired lady peered out.

Donna Buckley swung the door the rest of the way open. Her dimpled checks smiled and she threw her arms open wide. "Well, where's my half of your Nobel Prize?" She had retained a full pout of a mouth with finely aged wrinkle lines around the perimeter. She was pert in the hard-ironed style. Greg realized she would now be in her 80s. He embraced her.

It was well after midnight by the time Greg filled Donna in on his three trips in time, his relationship with Emily, and their flight from the director of the Time Agency. Donna unlocked an inner door to her heart upon hearing their plight. Greg explained their dilemma at not having personal money cards. Any attempt at getting one would attract immediate attention.

"After all, we're both over a hundred in 2035."

Donna said, "Let's talk more in the morning. Meanwhile you can have my spare bedroom."

"She's a wonderful old lady. Are we safe here?" Emily said.

"Yes, or I wouldn't have imposed on Donna. I think we can assume Webster kept the boys on a goose chase for the five years after Pismo and the search is over or floundering."

"How can we stay in this time, honey? Big Brother has arrived." Emily snuggled close in Greg's arms. "We can't even feed ourselves without that damned card."

"I'll try to locate Webster tomorrow to find out if Hoyer is still alive or in a position of power in 2035."

"Are you still committed to doing something about him?"

"If he's still in a position to hurt innocents, yes."

2030

TONY FRANCISCO was entering the director's office. Hoyer looked up and asked, "What did you find?"

"We unfortunately arrived a day after Philips' departure, and our computer control wouldn't allow us to start over again. We learned Gould had purchased a high voltage electron microscope from Hitachi, and it was installed in an abandoned storage room behind their High-grove house."

"Is it still there?"

"No."

"The son-of-a-bitch is traveling in time again. Go find him. He's making us look like a bunch of horse's asses."

2035

IN THE MORNING, GREG AND EMILY joined Donna in the kitchen of her two-bedroom bungalow. "Coffee?" Donna poured into a mug. "Breakfast will be ready in a minute."

"That would be great. Donna, do you have a newspaper where I can check stock prices?"

"Turn on the telly. Channel 70 will bring up the information menu."

Minutes later, Greg returned to the table and said, "Donna, we can pay our way around here." He turned to Emily. "Your stock changed names once, split three times, and increased five times in value since '68."

Donna looked at him over her plate with pensive eyes.

"Donna, Emily's going to sign over some of our stock certificates to you. In exchange, we'd like you to support us while we're here."

Donna appeared curiously indifferent. "That isn't necessary."

"Yes, it is. We can afford it. I took stock pricing information back with me in '53. It was better than inside trading. We bought and sold, not just low and high, but at the lowest and the highest. I think we're going to make you rich even by today's inflated prices."

Donna held her breath momentarily and then exhaled. Greg explained the need to find Webster. Donna took a last bite and then brought up the telephone directories for communities within commuting distance to NTSA, Virginia via the TV menu screen. Cody Webster's name didn't appear. She paged down to the status of unlisted names. Webster, C., appeared as a government-secured number available only to secret cleared government employees with ANTK (a need to know).

Greg said, "We have to find a way to contact Cody."

Donna said, "Well, let's think about it. In the meantime, we've got to get you two some modern clothes. You look hopelessly dated."

At mid-morning, they pulled into the Stanford Shopping Center parking lot. Donna was to visit a branch office of Merrill Lynch, where she would present the aged, but valid, stocks to be registered in her name and added to her PMC account. Her story to the broker was that she'd run across them recently in her attic years after she thought they were lost. They had been given to her by a family friend from Riverside ages ago. She would meet Greg and Emily back at the car in an hour.

Greg visited Macy's, decided on some clothes, and stepped onto a down escalator deep in thought about their predicament. On the one hand, it had been very fortuitous to find Donna and have their immediate necessities accommodated, but it was not a long-term solution. Donna could be their benefactor, but for how long? She was 80. In a year, five, or even ten, she could be gone and they would still have no access to their own PMC card. He had to make contact with Cody Webster to confirm that Hoyer's search had been abandoned.

"Daddy!" The scream yanked him out of his thoughts. He looked

toward the source of the scream. A tall, white-haired, old
lady was passing him on the other side of the escalator. She was staring
at him and so tightly griping the escalator rail, blood vessels stood out
on top of her age-spotted hand.

Chapter 35

ody Webster sat down across the desk from Tony Francisco.

C "What's up?" Cody's voice dripped with disrespect.

Director Hoyer had used the failed time search to demote Webster to the ranks and replace him with Tony Francisco as tracking administrator. The Congressional time committee had decided the search for Philips was to be abandoned. There had been no negative time splits in the five years of their real time between 2030 and 2035, and they concluded the revered father of time was either judicious or dead.

Francisco answered, "Something came across my desk an hour ago. I think it's worth us taking a look at, Cody. A couple of prospectors told the sheriff in Blythe, California, they'd come across a strange-looking metallic object the size of a room out in the Colorado Desert. It's probably a long shot, but I think you better check it out."

"Do you think it's a renegade escaping from future time? That would be a first for us."

Francisco shrugged. "I don't know. It would be a shocker though, if somehow Dr. Philips had succeeded in coming into our future."

"I thought the Los Alamos computer study proved he couldn't have single-handedly learned how to do that without access to the 1000 megabyte Intel computer chips of this century?"

Francisco toyed with his voice activated computer and glanced at its three-dimensional screen built into the opposing wall. "The Los Alamos computer was 99.5 percent confident he couldn't. That was good enough to keep us looking backward from 1968 for five years before Congress ordered us to stop. Nevertheless, you better check this out. See if it's anything other than another 20th century lost aircraft, and we've got a couple of old geezers with over-active imaginations."

Webster asked, "Have you told the director?"

Francisco turned away from the screen and looked at Webster. "No, it's such a long shot I don't want to get him on our case again."

"I'll leave in the morning." Webster ignored Francisco's attempt at comradery.

Webster walked through the maze of gleaming white tunnels, passed through a guard gate and climbed into his electric car. He sat for a moment behind the wheel and thought, I intended to resign back there but, instead, I've got to do one last favor for Philips. He had felt the seeds of discontent ever since returning to his time several months earlier and had decided it was time to make a change in his life.

He expected this trip to be as uneventful as the last five years. He'd return at the end of the week and submit his resignation. Maybe head home to Iowa, look up an old high school flame ... .

Chapter 36

reg exclaimed, "Oh my God, Carrie!"

G Greg embraced his dazed daughter. They stumbled off the escalator. Carrie looked up in disbelief, "Daddy, you're dead." Greg took her hand. "Take me to your car, Carrie."

"Are you a ghost?" Carrie's composure was return-ing.

As they approached her electric convertible two-seater Greg smiled and said, "No, honey. Please wait here a minute and talk to no one. I'll be right back."

He met Donna and Emily and told them about Carrie. He asked Donna to pick out a small wardrobe in his size and then he hurried back to Carrie. She stood waiting at her car, gripping the door handle. Greg put his hand on her shoulder and a surge of love swept over him.

"How are you, honey?"

"Daddy, you're dead! How can you be here?"

"Let's talk."

On her drive home, Greg briefly filled her in on the highlights of his life since leaving Menlo Park in 1983. She pulled into the driveway of their old home in Menlo Park. Once inside the house, she saw her father looking around. "Mom died in 2010 at 75. She was a bitter, old woman, daddy. When you were suddenly gone in 1983, we thought you'd committed suicide. That's when the guilt really hit me, and I started thinking about our family objectively. I remembered you had never responded to mother's detached , 'better than thou', treatment on you."

Greg looked around the living room, saddened that Helen had died so many years ago. "When did you find out I hadn't done myself in?"

"It was several years before the news of your trip into time shocked the world, and you became the most famous person on earth overnight." Carrie motioned her father to sit in an easy chair as she settled onto a couch.

"How did your mother take it?"

Carrie became thoughtful. "She was too proud to admit it, but I think she regretted the way she'd treated you."

While Carrie updated her father on her life the last 52 years, they moved into the backyard. He could smell the scent of gardenias that might have been the descendants of flowers he planted so many years ago with Pee Bee always at his heel. The edge of the Bermuda grass contained scarlet bougainvillea and hydrangea bushes.

She remembered, with crystal clarity his last goodbye and the inconsolable grief that had been mirrored in his eyes. It had haunted her until today. There were succulent tales of meeting her betrothed, now long dead of cancer in his 50s, and the birth of her only child, a son. She asked him about his parents she'd never met. With her mother, she'd visited her mother's family in Savannah and found them to be wealthy and materialistic. After nightfall, she flattened herself against the wall of the screened porch and stretched.

Greg said, "I think we've both had an exhausting day, honey. Can you drop me off on University in Palo Alto?"

"Of course, daddy."

"How was it?" Emily asked.

"Pretty emotional for both of us."

"Where did you leave things with her?"

"I'm going to meet her again for breakfast in the morning in the Stanford Mall coffee shop. I left it that it was imperative she not tell anyone I'm here."

"Good luck with that."

In the morning, Donna dropped Greg off at the coffee shop. As he entered, he saw Carrie rise from a booth facing the door and wave to him. A lean, gangly man with brown searching eyes, in his early 40s, stood up almost at military attention. Carrie reached out and hugged her father.

"Daddy, this is your grandson, Peter."

Greg looked down at her. "Carrie, I asked you not to tell anyone I was here."

"I didn't; just your grandson. I knew you wouldn't mind."

Peter stepped forward and extended his hand. Greg thought furiously for a moment and then shook hands. Peter studied his grandfather as he settled into the booth.

Carrie volunteered, "Peter received a doctorate from Stanford in physics."

"That's great! Where do you work?"

Peter hesitated while a waitress took their orders.
"I'm in the same building you were when you first studied Astral Projections and devised the EM tests to prove Bohm's holographic theory."

"You're at SRI," Greg commented.

"It was renamed after you a few years ago. I work at Philip's Research Institute. May I ask you where you've been all this time? What you've been doing? The world is going to be astounded and thrilled when it finds you're here."

"On the contrary. I'm a dead man if the word gets out. He looked at Carrie. "That's why I asked you to not tell anyone last night."

After breakfast, Peter drove them back to his mother's house contemplating on what he'd heard. He said, "So the Time and Space Agency director has severely prejudiced both Congress and the president against you, and there's a time manhunt for you."

"That's right."

"And Webster is your only friend within the system in 2035, but you can't find him because he's a government classified ANTK person."

"Yep."

"I have ANTK authorization from my position at Philips Research. I'll find Webster for you. Mother, do you realize how serious this is?"

"I do now." Carrie put an arm around each of them. "My, you are so much like your grandfather."

Chapter 37

G reg hugged Donna goodbye, turned and helped Emily board the high speed train. They entered their compartment and waved at Donna through the window. The train pulled out at dusk on a Tuesday evening.

Peter had obtained Cody Webster's personal telephone number and had reached Cody's voice recording saying he'd return in 24 hours. Greg was anxious to be in Washington when he returned. Donna purchased their tickets for a sleeping compartment on the San Francisco to Washington levitated train. She also made prepaid reservations for them at the Washington Marriot with arrangements for them to charge food and any other necessities to her PMC card during their stay.

They awakened in the morning as the train shot across the old Mississippi south of St. Louis. By the time they'd finished breakfast, they were flashing past Louisville, Kentucky.

"Beats airplane travel anytime," Greg thought.

After check-in at the Marriot, Greg tried calling Cody Webster from their room. He got the recording again and decided to try later in the evening. They walked out into the cool Washington afternoon and, like millions of tourists before them, walked for hours around the city, past the monuments and historical buildings and ended up at the Smithsonian Institute. They were both interested in seeing the Time Travel exhibit.

Greg's hair on the back of his neck bristled as he looked through the glass case at his fading handwriting on the open notebook he'd used during the months he worked with Ken Hoard.

By the time they exited, dusk was coming on and the streets were empty of foot traffic except for the homeless. Greg took Emily's arm and hurried them back to the safety of the Marriot. Washington at night looked like a besieged city.

As they entered the Marriot lobby, Emily pulled on Greg's arm. "I want out of this future, Greg. I hate it!"

"Me, too." Greg picked up the phone. "Let me try Cody again."

"Webster."

"Cody! Do you recognize my voice?"

"Oh my God! Where are you?"

Greg answered, "At the downtown Marriot."

"Downtown Washington?

"Room 925. Can you come over?"

"Be there in 20 minutes."

Greg and Emily used the 20 minutes to freshen up. Greg was drying his hands when he heard the knock at the door.

CARRIE AND PETER WERE clearing the dinner dishes when the doorbell rang. Two uniformed men stood on the steps: one about 30, the other, bigger, balding and middle-aged.

"Yes, can I help you?" Carrie asked.

The big man shoved her aside as Peter entered the living room.

"Oh, I'm glad to see the whole family's here," the big man said. "Hello, Peter,"

"What in hell do you think you're doing?" Peter's face was flushed.

"Why, one of you is about to tell me where I can find your grandfather," the man said and backhanded Peter. Carrie gasped as Peter fell to the floor. Hoyer turned to her. "We'll start with you, Carrie."

CODY WEBSTER STOOD before Greg and Emily. He was a little thicker in the waist and his hairline had started to recede. They started to shake hands for a moment and then bear hugged.

"So it was you."

"What do you mean?" Greg asked.

"I just returned from the Colorado Desert where a couple of prospectors found a peculiar machine. It was a primitive time machine."

"Damn! Does anyone else know?"

"I had to report in. Yes, my boss Francisco and the director both know."

Greg slumped into a chair. "There goes our escape out of here. Does Hoyer know it's us?"

"I don't know. I didn't talk to him directly. Francisco's thinking is leaning toward an escapee from time in our future. But that would be a first."

"What happens next?"

"It's being picked up by helicopter and will be taken to Victorville for inspection."

"They're going to find a modified 1968 Hitachi EM as the guts of it."

"Then they'll know it's you. Why did you come here? You were safe. We spent four months each year for five years looking for you until Congress told us to stop. I was so happy for you. Why in hell did you come here?"

Greg paused. "Hoyer killed my mother and friends when out of his time with impunity. He has to be stopped. I've come to stop him."

"You've come to kill the director?" Webster was incredulously.

"That was my plan and then escape into another future."

"Damnit, Philips! You couldn't have gotten past NTSA security to get to Hoyer." Webster was impassioned and stated with finality, "Now you're screwed!"

"Don't count me out yet, Cody."

"Why? Are you still going to dance on his grave?" Webster

smirked.

"Did you keep my 1983 letter to Southwell, the newspaper personal, and the film?"

"Yes."

"Where is it?"

"I returned it to the Time Travel Memorial in La Honda. I figured it had been safe there for 40 years. Like a genie. It's back in the bottle."

"Then maybe Emily and I have one option left. If I can get to Hoyer before the Victorville inspection of the machine identifies it as mine."

"And?"

"And let Hoyer know I've sequestered highly incriminating evidence against him that also exonerates me of any wrong doing. I'll make a trade. I'll leave the evidence hidden in 2035 and not expose him if he returns my machine and lets us get the hell outta here."

"Give up the idea of killing him?"

"Do you see that I have a choice?"

"No, I'll help you as long as I'm not an accomplice to a murder. We've got to move quickly."

Webster's phone rang. Ten minutes later, Webster hung up the phone grim faced. "Bad news. That was Francisco calling from Menlo Park. The moment Hoyer heard about the machine, he guessed it was you and took Francisco with him to Menlo Park. He has your daughter and grandson under house arrest."

Greg sat bolt upright. "Can he arrest them without a charge? Where the hell is the Constitution in this century?"

"No, of course he can't legally, and it's got Francisco spooked. He said Hoyer was almost deranged in apparent fear — beyond reason. Francisco thinks he might kill them."

"You know damn well he will! He'll interrogate them and kill them! That's his modus operandi!"

"He's already interrogated them. Your grandson confessed after Hoyer knocked your daughter around. He admitted seeing you, but didn't know your whereabouts."

Greg asked, "Are they safe with Francisco there?"

"For the moment. Hoyer's been called back to Washington. He told Francisco you were here to kill him. Francisco argued you had no reason, but was ignored. He thinks Hoyer is holding your relatives hostage for his own security until you're caught."

Greg took a deep breath. "My options are closing."

He gazed across the room at Emily sitting tensely on the bed.

"Cody, will you take Emily back to 1968?"

"Okay. The computer control should still allow time travel to 1968."

Emily shouted, "No! I don't go anywhere without you!"

Greg slowly raised out of his chair and walked across the room, put both hands on her shoulders and looked intently into her eyes. "Honey, I've been on borrowed time ever since returning to 1983. Cody's right. My plan was destined to fail. All I've accomplished is to put Carrie and Peter at risk and put you in a time you don't belong. The machine is gone. I'm stuck in 2035."

"Why can't Cody take both of us home to 1968?" She squeezed her hands together.

"You know that's hopeless. Hoyer knows I'm here, and he'll know if Cody takes me anywhere."

"The same thing applies to me," Emily replied stubbornly.

"You're no threat to Hoyer, and Cody will just be doing his job by returning you to your own time."

Emily looked into the depth of his hazel eyes and her own filled with tears. "I don't care how impractical it sounds. I'm staying with you."

Greg shook his head sadly and said slowly. "I'm going to offer myself as a trade for Carrie and Peter."

"No!" Emily stared at him in horror. After a long moment she whispered, "You can't trust him to honor a trade."

"I'll hold him accountable with the incriminating evidence. It's the only choice I have."

Emily wilted and nodded her head.

Greg turned to Webster, "How do I get into NTSA to find Hoyer?"

"I'll get you in with my fingerprint pass card. I'll go with you in the morning and then take Emily to Goddard and back to 1968. Oh, and I'd better draw you a map to get to the director's office. The place is a real maze."

Greg and Emily spent another night in each others arms. She quietly sobbed before finally falling asleep. She dreamed she was plummeting down a slippery slope and found a little horn of ice to grasp. When she attempted to embrace it, the projection broke off.

At daybreak, they quietly talked in halting terms about the ephemeral nature of life. They sadly rejoiced for the months they'd had together. Greg explained his conviction that life... existence, is not as it seems...as we perceive it in our little world of time, space, and shapes. There is a just and living God that would unite them again in another place. They kissed for the last time in the NTSA parking lot in the morning. Emily said simply, "I'll wait for you in my time until I die." She waited in her own black and bitter place for the ride to Goddard and her flight home.

Webster passed his card and hand palm down across the bar reader at the entrance to NTSA and a heavy electrically controlled gate unlatched to allow one person to pass. Greg walked through, dressed in his most conservative 2035 apparel. He passed among the sandstone mountains interlaced with tunnel passageways that was the home to the Time and Space Agency. He passed employees dressed in cerulean blue uniforms in the hallways without incident and consulted his hand drawn map several times before finding his way to the director's inner-sanctum. He pushed open the massive door and entered the reception area. "I'm here to see Director Hoyer."

The attractive brunette looked him over and said, "I don't believe you have an appointment."

"He'll see me."

"Who shall I say it is?"

Greg gave her a tight-lipped smile. "Tell him it's the tough guy."

Chapter 38

Herman Hoyer stood hunched forward, hands pressed firmly on the top of his desk for support. Greg blinked at the metamorphose of the man. His promontory of hair was gone, replaced by a mole-mottled scalp. He said into an unexposed intercom, "Take the afternoon off, Elizabeth."

"But....?"

"Take it off."

"But you have a joint appointment at three with the committee and the president."

"I'll keep it. Get out of here."

Greg said, "Herman. You look like hell."

Hoyer turned to Greg. "Why are you here?"

"To make a deal."

"A deal? You're on my turf, unarmed, and you're talking deal?

"You don't know I'm unarmed." Greg shifted his weight in the uncomfortable chair.

"Your body was scanned when you passed Elizabeth's desk. If you'd been armed, my door would have automatically been dead bolted."

Hoyer's right hand momentarily dipped from view and re-emerged with the hand weapon he'd pointed at Greg in 1953. His left hand reached for the phone.

"You'd better hold up on that, if you don't want us ending up in adjacent cells," Greg threatened.

Hoyer's hand fluttered above the phone.

Greg said, "I've got proof you murdered Ken and Dave Hoard and a CIA agent in 1983," Greg said. "You've got my daughter and grandson. I'm here to trade myself for them."

"Go on."

"I left a note out for you the night I traveled back to 1968. It was a ruse. I left another for the CIA agent Southwell explaining you murdered Ken and Dave. It turns out you murdered him, as well. I have that note and my 1953 newspaper personal to Ken Hoard. And I've got them here in 2035. If you don't accept my trade they will automatically be delivered to The Washington Post, and you're career is over."

Hoyer said, "You don't have anything for me to worry about."

"Oh, and I also have photographs Dave Hoard took of me as I departed 1983 for 1953." Greg calmly recrossed his legs.

"Photos?"

"Photos that show me in the machine departing from 1983 for 1953, including the time and date, taken by the Hoards."

"So you believe if you go public with this stuff it will cast suspicion on me?"

"Yep, Herman, that's what I think."

"You want to trade yourself for your family's release?"

Greg nodded.

"You have nothing. I've given you a lot of thought. My first idea was to send you off 10,000 years into the past or 10,000 years into the future. But, with time, my ideas became more sophisticated. I decided to put you back in your machine and send you off without your ion gun functioning. You would be forever lost in time in a black hole." Hoyer paused and smiled. "But, I finally came up with the most exquisite solution. By the end of the day, you will have never existed, and of course neither will your daughter and grandson, or this evidence of yours."

Greg stiffened.

Hoyer said, "You wonder how can that be? You think you're a

fundamental requisite for us in the 21st century? Sorry. I've had Los Alamos conduct a computer simulation of a time split caused by your mother aborting you in 1934."

"It seems the world wouldn't have missed a beat. Any one of a number of scientists with 200 KV electron microscopes would have found a mini-black hole and time traveled very quickly thereafter."

"You're going to cause my abortion before I was born?" Greg hoarsely asked. The dreadful finality of the threat hitting him in the solar plexus like a physical blow.

"The president knows you are a homicidal maniac in time."

"He's considering your proposal?"

"He's approved it. I meet with him for rubber-stamp approval at 3:00 this afternoon. All he wants to see is my report on the Los Alamos simulation findings."

"You bastard!"

The room suddenly started to turn and Hoyer's lips were moving but his raspy voice was only faintly coming across a vast canyon to Greg. The room had a hazy ephemeral ambiance to it as Hoyer's voice droned on. Greg gripped the metal chair's arms, as if to hold on from spinning out of control. He watched superimposed on the contents of the room amorphous images of Ken, Dave and Southwell. Ralph and he, as boys and as men. His mother walking toward him swinging a game leg.

They became more defined and collectively turned to face him, silently crying out for revenge against the man who had destroyed their lives. Greg's vision cleared and he could see Hoyer's lips moving beneath wild, insane eyes. The hollow voice droned on. "So I'm going to leave here in a few minutes for my meeting with the president. I'll go directly from the meeting to 1934 and abort your fetus. You won't be here when I return because you will have never existed."

Hoyer walked around the desk. "But first, I'm going give you some pain, tough guy." He thrust the gun forward. It roared and Greg's knee cap exploded. He gasped in horrible pain as Hoyer aimed again and fired. His second knee cap exploded. From the empty reception area, two more muffled shots could be heard if anyone had been present.

Chapter 39

RIVERSIDE - 1968

E mily locked her front door and started her brisk three-mile walk to school. It was the 23rd such walk since Webster had returned her home.

She had been hopeful the first few days after her return that somehow Greg would perform another one of his miracles and again prevail over Hoyer. With each passing day her hopes dwindled. Nevertheless, she had decided not to drive to and from school. She was superstitious about the daily walk. She had been on the walk when she first met the adult Greg and began the happiest time in her life.

She had continued for 15 years after Greg's capture and return to Menlo Park, even though she had a car by then. The walk was like a medieval talisman for her. It had brought her lover to her two times before and might again.

It crossed her mind that another Greg existed in 1968. A much younger Greg working at SRI in Menlo Park and living with Helen. She considered calling him. But what would she say? He would know nothing about his time travel in the future and would be devotedly raising a teenage daughter. What would she say if she called him? "This is your old school teacher, Emily. You're going to become the father of time travel in 15 years and will return to 1953 and fall in love with me." She looked at this possibility from every angle and was discouraged.

"He'll think I've lost my marbles. He's in an entirely different life in 1968 and an entirely different frame of mind."

"I'll call, nevertheless, as a last resort. Or maybe show up at SRI and explain it in person?" It wasn't going to work, she knew, but she would have to try. "I'll give him another month to come for me. But what if Hoyer has killed him in 2035?"

She walked on toward the arroyo. She could make out two men standing in the distance looking out over the football field below. One of them appeared to be an emaciated, white-haired, old man supporting himself on crutches.

Chapter 40

2035

Greg woke up in an intensive care ward in the Bethesda Naval Hospital. An orderly noticed and called the doctor. Within minutes, there was a retinue of doctors in the room. Greg tried to speak but was asked to wait. He looked down at his body and found his left arm and both legs elevated in casts. He was being fed intravenously. He looked around the room and saw Cody Webster lurking in the doorway.

The doctors held a brief caucus at the foot of his bed. They looked as a group over at Greg and smiled. They were pleased about something. They exited in single file and nodded Webster in. "Hi, Dr. Philips. Don't answer. You're not supposed to talk. You're very weak and still not out of the woods. You've been in a coma for a week. I'm only allowed to stay with you for a few minutes."

"What happened," Greg whispered.

"I first learned you'd killed the director late in the afternoon after I'd dropped you off at NTSA. I got a call from Francisco telling me to meet him at the Pennsylvania Avenue entrance to the White House immediately. When I arrived, we were ushered in to President Jerry Cole and several Congressional members of the time travel committee.

"I was instructed to leave immediately for the year 1934. I was given a small bottle of liquid to administer to your mother at my discretion. It would cause her to abort your fetus within minutes. I asked what was going on, and the president took charge and told me you had successfully penetrated NTSA and murdered the director. By executive order, he had mandated your immediate abortion. I asked where you were at the moment, and he told me you were in an intensive care unit here. That, unfortunately, the agents who found you unconscious on the floor of the director's office had called for an ambulance and saved you.

"I interrupted and told them your whole story. They couldn't believe it. Hoyer had that much prestige with them. I told them of your proof in La Honda and President Cole himself placed a call directly to the Memorial. The rangers quickly emptied the old aspirin bottle and within the hour the film had been processed. The prints were faxed, along with the newspaper personal and letter, directly to the White House."

"How bad off am I?" Greg interrupted.

"You had a bullet within millimeters of your heart that required a serious operation. Luckily, it had first struck the bones in your left arm and decelerated," Cody explained. " And Hoyer shot off both of your knee caps. You're not strong enough yet for the doctors to start reconstruction surgery."

Greg's memory returned of his last moments in the director's office. Before he could get out of his chair, Hoyer had blown away his left knee cap, aimed again and blown off his right. Relishing the moment, Hoyer had walked slowly over and shoved his gun directly into Greg's crotch. But before he fired, he leaned over and looked into Greg's pained eyes. It was his moment of triumph. It killed him.

Greg jammed the stiffened fingers of his right hand into Hoyer's Adam's apple and grabbed the gun barrel with his left and twisted. Hoyer managed to pull the trigger again as Greg twisted the gun away from his crotch. The bullet crashed through Greg's left arm and ripped into his chest. Hoyer was gurgling, his face a frightened mask of bloated red. Greg, with the last of his strength, used his right hand to twist the gun slowly into Hoyer's face. A look of terror crossed

Hoyer's face as he gagged. After the longest moment in his life, the bullet tore into Hoyer's mouth and ripped off the back of his head. Greg's last memory as he blacked out was seeing the diaphanous figures of Ken, David, Jeff, Ralph, and his mother and father standing across the room nodding approval at the bloody scene.

In the weeks that followed, Greg's strength increased and the reconstructive operations on his knees commenced. He spent the days lying in his hospital bed, watching unfolding news stories about himself. The world had been electrified with the news that the father of time travel was alive in 2035, Carrie and Peter were instant celebrities and moved into the Washington downtown Marriot and visited Greg daily. Donna Buckley became an overnight heroine and was given a lucrative advance for her life story, which covered her role in assisting Greg in his mini black hole experiments right through sheltering him in the weeks preceding his confrontation with the director. Cody Webster received presidential approbation for heroism above and beyond the call of duty.

As the story unfolded, Greg's love affair with Emily Johnson captured the world's imagination and a movie of their love was to be filmed. The current leading ladies of the cinema all coveted the role of Emily. The president of the United States visited Greg several times over the weeks and they became good friends. At the end of the fourth week, he strolled beside Greg as he tried out his new crutches around the hospital grounds. President Cole was about Greg's height and weight and a dozen years older with a thatch of white, fashionably groomed hair. He was coming to the end of his second term.

"You've sure turned the world on its ear, Greg. Not that I'm complaining. Hell, my popularity ratings have never been higher, but we've got a problem."

"What's that?"

"What to do with you. Los Alamos simulations point out the potential time split dangers if you stay in our time. I've discussed the problem with the Congressional committee, and we're in agreement. You can stay if you want, or we can have Webster take you back to your own time. We owe so much to you; it can be your choice."

"There's nothing for me here or in my own time. I want to go back to Emily in 1968."

"We were afraid you would want that."

Greg smiled. "I was in 1968 for quite a spell and didn't cause any major time disruptions. I'm pretty experienced at avoiding time splits. I just want to live out my time with Emily."

The president said, "You can go whenever you're ready."

The Bethesda Naval Hospital surgeons wanted to keep him undergoing physical therapy for several additional months. He thanked them and declined. Carrie and Peter visited him before his departure for a last goodbye. The following morning, Cody showed up at noon and they departed for Goddard.

1968

Emily stopped, her left hand holding the arroyo railing for support.

The white-haired man leaned forward on his crutches and stared in her direction. He said something to the younger man. Was the younger man Webster? Her heart almost stopped. The white-haired man turned away from the younger man and began moving toward her swinging his body forward on crutches. She stood transfixed, unable to move. As he got closer, she could make out a squinting, all-out grin covering his face.

Emily dropped her briefcase and ran to him.

THE END